SpringerBriefs in Water and Technology

More information about this series at http://www.springer.com/series/11214

Gholamreza Asadollahfardi

Water Quality Management

Assessment and Interpretation

Springer

Gholamreza Asadollahfardi
Civil Engineering Department
Kharazmi University
Tehran
Iran

ISSN 2194-7244 ISSN 2194-7252 (electronic)
ISBN 978-3-662-44724-6 ISBN 978-3-662-44725-3 (eBook)
DOI 10.1007/978-3-662-44725-3

Library of Congress Control Number: 2014948748

Springer Heidelberg New York Dordrecht London

© The Author(s) 2015

This work is subject to copyright. All rights are reserved by the Publisher, whether the whole or part of the material is concerned, specifically the rights of translation, reprinting, reuse of illustrations, recitation, broadcasting, reproduction on microfilms or in any other physical way, and transmission or information storage and retrieval, electronic adaptation, computer software, or by similar or dissimilar methodology now known or hereafter developed. Exempted from this legal reservation are brief excerpts in connection with reviews or scholarly analysis or material supplied specifically for the purpose of being entered and executed on a computer system, for exclusive use by the purchaser of the work. Duplication of this publication or parts thereof is permitted only under the provisions of the Copyright Law of the Publisher's location, in its current version, and permission for use must always be obtained from Springer. Permissions for use may be obtained through RightsLink at the Copyright Clearance Center. Violations are liable to prosecution under the respective Copyright Law.
The use of general descriptive names, registered names, trademarks, service marks, etc. in this publication does not imply, even in the absence of a specific statement, that such names are exempt from the relevant protective laws and regulations and therefore free for general use.
While the advice and information in this book are believed to be true and accurate at the date of publication, neither the authors nor the editors nor the publisher can accept any legal responsibility for any errors or omissions that may be made. The publisher makes no warranty, express or implied, with respect to the material contained herein.

Printed on acid-free paper

Springer is part of Springer Science+Business Media (www.springer.com)

Acknowledgments

I would like to express my gratitude to Mr. Mohsen Asadi and Mehrdad Amani for their help in the preparation style of the book and for English correction.

Contents

1 Introduction 1
   1.1   Introduction 1
   References 3

2 Selection of Water Quality Monitoring Stations 5
   2.1   Historical Background 5
   2.2   Sanders Method 6
   2.3   Multiple-Criteria Decision Making (MCDM) Method 8
      2.3.1   Making Dimensionless 9
      2.3.2   Assessment of Weighting (Wj) for Attributes 10
   2.4   Dynamic Programming Approach (DPA) Method 10
      2.4.1   The DPA Theory 10
      2.4.2   Normalization and Uniformization Procedure 12
   2.5   Application of Sanders Method 14
      2.5.1   Comment on the Application 19
   References 19

3 Water Quality Indices (WQI) 21
   3.1   Historical Background 21
   3.2   Summary of WQI Methods 26
   3.3   National Sanitation Foundation's Water Quality Index
      (NSFWQI) Method 26
   3.4   British Colombia Water Quality Index as (BCWQI)
      Method 29
   3.5   Application of NFSWQI Method, a Case Study: Kārūn
      River (Iran) 30
   3.6   Application of NSFWQI Method in Sefid-Rud River (Iran) ... 33
   3.7   Application of BCWQI Method in Sefid-Rud River (Iran) ... 36
   3.8   Comments on Application 38
   References 38

4	**Time Series Modeling**		41
	4.1	Introduction	41
	4.2	Historical Background	42
	4.3	Time Series	44
	4.4	Forecast Error	46
	4.5	Box-Jenkins Methodology for Time Series Modeling	47
	4.6	Stationary and Non-stationary Time Series	48
	4.7	The Sample Autocorrelation and Partial Autocorrelation Functions	49
	4.8	Classification of Non-seasonal Time Series Models	51
	4.9	Guidelines for Choosing a Non-seasonal Models	52
	4.10	Seasonal Box-Jenkins Models	52
	4.11	Guidelines for Identification of Seasonal Models	54
	4.12	Diagnostic Checking	54
	4.13	Exponential Smoothing Methods	56
		4.13.1 Simple Exponential Smoothing	57
	4.14	Winter's Method	58
	4.15	One and Two-Parameter Double Exponential Smoothing	59
	4.16	Adaptive Control Procedures	60
	4.17	Application of Time Series	61
		4.17.1 A Case Study: Latian Dam Water Quality	61
	4.18	Summary	75
	References		75
5	**Artificial Neural Network**		77
	5.1	Introduction	77
	5.2	Historical Background	78
	5.3	Artificial Neural Network Theory	79
		5.3.1 Theory of ANN	79
		5.3.2 Dynamic ANN Models	81
		5.3.3 Data Preparation	81
		5.3.4 Learning Rate	82
		5.3.5 Model Efficiency	82
	5.4	Application of Artificial Neural Network	83
		5.4.1 A Case Study: Zaribar Water Quality (Iran)	83
	References		90
6	**Introducing of Ce-Qual-W2 Model and Its Application**		93
	6.1	Introduction	93
	6.2	Historical Background	94
	6.3	Theory of the Model	95
		6.3.1 Boundary Conditions	98

6.4	Study Area		98
	6.4.1	The First Case Study	98
	6.4.2	Second Case Study	111
References			117

Chapter 1
Introduction

Abstract The abatement of fresh water and groundwater resources, increasing population as well as industrial demands render the protection and preservation of these resources all the more important. Water quality is a term applied to indicate the suitability or unsuitability of water for various uses. Each type of water uses needs certain physical, chemical and biological characteristic while various uses have some common characteristic. Water quality management is the management of water quality of the physical, chemical and biological characteristic of water; therefore, management and regulatory agencies can use to evaluate alternatives and make necessary decisions. In this chapter, after defining a few water quality terms and a brief review of the significance of water quality management, the framework of the book was described.

1.1 Introduction

Life on this planet is positively linked to water. The dwindling fresh water resources, increasing population as well as industrial demands render the protection and preservation of these resources all the more important. The situation is particularly of concern in semiarid and arid countries with growing population and industry. Groundwater resources are scanty and the rainfall, though meager, is not uniform. One such country is Iran, located in the Middle East, to receive merely enough rainfall in its Northern and Western parts to sustain those regions themselves, while the rest of the land mass is either semi arid or arid. Hence, healthy sized budgets and time need to be devoted to constructing infrastructure for the transportation of drinking water from the more endowed parts of drier ones. Needless to say, protection and maintenance of water quality over the long haul is very important.

It is necessary to establish the current quality of surface and ground water resources before measures can be taken to control water pollution. Briefly, the process involves establishing water quality monitoring stations along the waterway to collect samples for the analysis of the characteristic of water followed by rigorous interpretation of the collected data since the colossal amount of data without proper interpretation can, in no way, lend any effective aid to water quality management. Many methods have been developed to interpret the data. Deterministic and statistical methods are examples. However, among all the methods available today, water quality indices are perhaps among the simplest and yet extremely useful methods. The results of interpretations derived from this method are understandable for both water quality managers and the general public.

Water quality is a term applied to indicate the suitability of water for various uses. Each type of water uses needs certain physical, chemical and biological characteristic while various uses have some common characteristic. The composition of surface and ground water depends on the characteristic of the catchment area such as geological, topographical, meteorological and biological of the area. Water quality in various areas is hardly ever constant, and the variations are caused by changes in concentration of any inputs to water body. Such variations may be natural or man—made and either cyclic or random. Random variation of water owing to unpredictable events. As an example, storm can increase flow and increasing pollution due to the wash of its catchment area. Therefore, the nature of water quality is stochastic and deterministic. Consequently, for proper interpretation of the data understating both of the characteristic is vital.

Water quality monitoring is the effort to find quantitative information on the physical. Chemical and biological characteristics of water using statistical sample (Sander 1983). Monitoring means watching the ongoing flow of water to make sure no law and regulation are broken. However, the word has a different meaning when utilized to refer to water quality measurements, as a result has the term network taken on a meaning beyond the strict definition of the word when referring to water quality monitoring. Network design means to determine the location of sampling stations (Sander 1983). The location of sampling stations and type of water quality parameters depends on the objective of water usage. The water quality situation is a function of complex natural and man-made causes and of the resulting integration in both space and time. Therefore, abstracting the core of the water quality conditions at a reasonable cost is very difficult.

For water pollution control, it is necessary to figure out surface and ground water quality. The first stage in this process is to establish water quality monitoring stations to collect samples to analyze the characteristic of water. The second stage is type of sample collection which is very important because collected samples should be representative of the water body. The third stage involves interpreting the collected data since huge amounts of data without proper interpretation cannot help water quality management effectively (Asadollahfardi 2000).

The first stage in water quality management is establishing enough and suitable selected monitoring stations considering the objective of water uses. The Second stage is the availability of enough data with proper precision regarding the aim of water use. The third stage is an interpretation of the data which the outcome of this step can help water quality management for water quality planning to control water quality.

The principal aim of the global freshwater quality monitoring project, Global Environment Monitoring System (GEMS)/WATER presents a descriptive example of the intricacy of the assessment task and its relation to management (WHO 1987), to offer water quality assessments to governments, the scientific society and the public, on the quality of the world's fresh water in relation to human and aquatic ecosystem health, and global environmental concerns, specifically:

- To describe the rank of water quality;
- To spot and quantify trends in water quality;
- To define the cause of observed conditions and trends
- To identify the types of water quality difficulty that happen in specific geographical areas; and
- To provide the accumulated information and assessments in a form that is for resource

In other words, water quality management is the management of water quality of the physical, chemical and biological characteristic of water (Sanders 1983); therefore, management and regulatory agencies can use to evaluate alternatives and make necessary decisions.

This book consists of six chapters. This chapter includes significance and definition a few terms of water, and necessary steps for water quality planning; in Chap. 2, in the first step the significance of appropriate water quality site selection is defined and then a summary of Sanders method, Multiple-Criteria Decision Making (MCDM) and Dynamic Programming Approach (DPA) is described and finally, an application of the Sanders method for existing water quality monitoring stations in the Kārūn River is assessed. Chapter 3 encompasses the previous researchers' work, detailed information of National Sanitation Foundation Water Quality Index (NSFWQI) and British Columbia WQI and two case studies using NSFWQI and British Columbia WQI. In Chap. 4, the historical background of using time series, a summary of Box-Jenkins time series and method of building, diagnostic and predicting the future of the time series model, as well as a brief explanation of exponential smoothing and the Winter's method is described. Finally, as a practical exercise, an application of time series model as a case study is depicted. Eventually in Chap. 5 a summary of artificial neural network and as a case study is discussed. In Chap. 6 introduces the deterministic model Ce-Qual-W2 and then two applications of the model are described. The first application is to study changing of total dissolved solids in Karkheh Dam in southwest of Iran and the second study is about Kārūn River in Khouzestan province, Iran.

References

Asadollahfardi G (2000) A mathematical and experimental study on the surface water quality in Tehran. Ph.D. Thesis, University of London, London

Sanders TG (1983) Design for networks for monitoring water quality. Water Resources Publication, Littleton

WHO (1987) GEMS/WATER operational guide. World Health Organization, Geneva, Switzerland

Chapter 2
Selection of Water Quality Monitoring Stations

Abstract Due to financial constraints and improper selection of water quality stations considering the objective of water uses, water quality monitoring network design is an efficient method to manage water quality. The most crucial part is to find appropriate locations for monitoring stations. In the past, most of water quality selection stations were subjective and the designs on the network had some human error. However, now there are several mathematical methods using experimental data for assessment of existing monitoring stations or designing new network such as Sanders method, multiple criteria decision making (MCDM) and dynamic programming approach (DPA) that developed by researchers. In the following chapter, after reviewing the historical background of developing and application of the methods, the theory of these methods was described in details. Finally, the application of the Sanders Method to design number of water quality monitoring stations in the Kārūn River which located in the south west of Iran was studied.

2.1 Historical Background

Allocation of the water quality monitoring site is the first and the significant step in the design of the water quality network. The importance of water quality network control concerning pollution causes creation of water quality stations in the network. However, finical constraints causes to decrease the number of water quality station in the network. Regarding optimizations of a number of monitoring stations some techniques were developed such as Sanders method, artificial neural network, Multi-Criteria Decision Method (MCDM) and Dynamic Program Approach (DPA). Some researches were carried out by Sharp (1970, 1971); Dandy (1976); Sanders (1983); Schilperoort and Groot (1983); Ward and Loftis (1986); MacKenzie et al. (1987); Woldt and Bogardi (1992); Harmancioglu and Alpaslan (1992); Hudak and Loaiciga (1993); Karemi (2002); Ozkul et al. (2003); Khalil and Quarda (2009); Noori et al. (2007); Karamouz et al. (2009); Khalil et al. (2011); Asadollahfardi et al. (2011). The DPA technique, which is a general method for maximizing and minimizing mathematical functions for solving a problem together with Multi-Criteria Decision, was introduced by Bellman (1957). Letternmaier et al. (1984) suggested an optimization method, which they used the DPA, for inspection of water quality station. They applied the technique for reduction of the number of stations in the urban water quality monitoring network. The results showed a reduction of monitoring stations from 81 to 47. The DPA was studied and extended by some researchers such as Harmancioglu et al. (1994, 2004) and Icaga (2005). Harmancioglu et al. (2004) applied the DPA on Gediz River in the West of Turkey for reduction of water quality monitoring stations. Icaga (2005) assessed existing water quality monitoring stations of the Gediz River applying the DPA for different water usage and allocated a different weight for indices of each water quality monitoring site. Cetinkaya and Harmancioglu (2012), applied the DPA for assessment of water quality monitoring stations. The results showed that the DPA was a suitable tool for optimization of the number of monitoring stations which are going to be remaining. Asadollahfardi et al. (2014) used the DPA for assessment of existing water quality of Sefīd-Rūd River in North of Iran. The results described that the DPA reduced the number of stations of the network.

As mentioned previously, there are numbers of methods for selection of stations. In this chapter Sanders, MCDM and dynamic approach is briefly described.

2.2 Sanders Method

The method proposed by Sanders et al. (1983) involves the identification of sampling reaches in a river basin (Macro location) when the intent is to allocate monitoring sites along the entire basin. According to Sanders et al. (1983), the objectives of the sampling must be defined prior to the actual design process.

The emphasis on water quality management efforts has recently been shifted from detection of stream standard violations to the assessment of overall trends in water quality because of various complications in compliance monitoring, such as intermittent or random sampling practices and incorrectly selected sampling locations. As a result, restrictions on effluent quality have become more significant than those on stream quality. In this case, a network developed for the assessment of trends must cover sampling points which will yield information characteristic of reaches of the river and in composite with other stations will yield information characteristic of the condition of the river system in general (Sanders et al. 1983). Sanders et al. (1983) proposed their method for site selection in a water quality monitoring network with the primary objective of detecting, isolating and identifying a source of pollution. Sanders et al. (1983) describe three approaches for macro location:

- Allocation by the number of contributing tributaries;
- Allocation by the number of pollutant discharges;
- Allocation by measures of BOD loadings.

These approaches, although they may produce a rather different system of stations, work pretty well in initiating a network when no data or very limited amounts of data are available. It must be noted that, by applying these methods, one may roughly

2.2 Sanders Method

specify the appropriate sampling sites. To pinpoint the locations more precisely, micro location and representative sampling considerations must be followed.

The first approach systematically locates the sampling sites so as to divide the river network into sections which are equal in respect to the number of contributing tributaries. The first stem involves stream ordering, where each exterior tributary or link contributing to the main stem of the river, (one which has no other tributaries or one with a certain minimum mean flow) is considered to be of the first order. Ordering is carried out along the entire river such that a section of the river formed by the intersection of two upstream tributaries will have an order described as the sum of the orders of the intersecting streams. At the mount of the river, the magnitude (order) of the final river section will be equal to the number of all contributing exterior tributaries.

Next, the river is divided into hierarchical sampling reach by defining centroids for each reach. The major centroid which divides the basin into two equal parts is found by dividing the magnitude of the final stretch of the river by two. Accordingly, the major centroid where a first hierarchy station is to be placed is located in that link whose magnitude is closest to:

$$M_{ij} = [(N_0 + 1)/2] \qquad (2.1)$$

where M_{ij} = the first hierarchical location; with M = the magnitude (order) of the link, i = the hierarchical level of the station to be placed on that link, and j = the order of that station within the ith hierarchical level (e.g., M_{11} = the first station at the first hierarchical level and M_{12} = the second station of the same hierarchy). N_0 = the total number of exterior tributaries at the most downstream point of the basin where station M_{11} is placed M_{12} (Or the stream number closest to it) divides the total catchment area into two equal parts for which new centroids may be determined.

In the above procedure, it must be noted that a link determined at a given hierarchy does not necessarily have the value of M_{ij} because a link of that number may not exist. In this case, the link closest in magnitude is chosen as the centroid when this link is specified a sampling location is placed at its downstream junction. Although Sanders et al. (1983) located the station M_{ij} at the downstream point of the reach that has the corresponding stream order number, it may be allocated to any site along that reach, considering such local factors as accessibility of the site or the presence of a stream gauging station also; note here that the squared brackets in Eq. (2.1) indicate a truncation of the enclosed value to an integer value.

As noted above, M_{12} divides the total basin into two equal parts where new centroids may be determined for the upstream part, the first station with the second hierarchical order is found by which is the magnitude of the link that divides the region upstream of M_{12} into two equal areas with respect to their drainage density.

$$M_{21} = [\frac{M_{12} + 1}{2}] \qquad (2.2)$$

Essentially Eq. (2.2) applies the same procedure as in Eq. (2.1) by replacing N_0 with M_{12}.

For the downstream portion of M_{12}, one can either renumber the tributaries, or alternatively, the centroid may be found as the location with an order closest to either:

$$M_{ij} = (M_d - M_u + 1)/2 \qquad (2.3)$$

$$M'_{ij} = M_u + M_{ij} \qquad (2.4)$$

where, i = the hierarchy order; j = the order of the station; M_d = the order where the basin is divided on the downstream side; M_u = the order where the basin is divided on the upstream side. This procedure locates the station at the second hierarchical level as M_{12} and M_{22}. So that's two more sampling locations are added to the system, which now has four stations altogether in the first and second hierarchical levels.

Next, new stations may be allocated upstream and downstream of both M_{21} and M_{22} to constitute stations at the third hierarchical levels. This is accomplished by applying the same procedure described in Eqs. (2.1–2.4). Eventually, four new locations will be designated at the third hierarchical level so that the network now comprises eight stations altogether.

Having specified third hierarchical stations, the same procedure is applied to select higher order hierarchy locations, if necessary. Here hierarchy levels indicate sampling priorities so that increasing hierarchies show decreased levels of sampling priorities. How far the hierarchical divisions should be continued depends on economic considerations and information expectations from sampling at each hierarchy.

In the second approach proposed by Sanders et al. (1983), the same procedure is used by cumulatively numbering the discharges from polluting sources as if they are exterior tributaries. Consequently, the sampling locations are determined as functions of populations and industrial activities. In both approaches, the sampling stations are to be placed at the downstream end of a river segment before an intersection.

2.3 Multiple-Criteria Decision Making (MCDM) Method

MCDM is applied in complicated decisions making processes. The differentiating feature of these methods is their application of more than one criterion. The models are divided into two main groups: First, Multiple Objective Decision Making (MODM) models and second, Multiple Attribute Decision Making models (MADM). MODM models are applied for design whereas MADM models are employed to select the best options. The MADM is defined by the following matrix (Table 2.1).

2.3 Multiple-Criteria Decision Making (MCDM) Method

Table 2.1 Matrix of MADM

	X_1	X_2	X_3	....	X_n
A_1	r_{11}	r_{12}	r_{13}	....	r_{1n}
A_2	r_{21}	r_{22}	r_{23}	....	r_{2n}
A_m	r_{m1}	r_{m2}	r_{m3}	....	r_{mn}

$A_1, A_2, \ldots, A_m$, in decision making matrix D, indicate m predetermined alternatives (Such as sampling stations in this work), $X_1, X_2, \ldots, X_n$ show n attributes (such as population, area of basin, water qualitative parameters, …) to assess desirability of each attribute. The members of matrix describe the special values of jth attribute for jth alternative. The optimal solution for a MADM consists of the most suitable assumed alternative A^*.

$$A^* = \{A_1^*, A_2^*, \ldots, A_n^*\}$$

$$X_j^* = \max i \; uj \, (r_{ij}) \tag{2.5}$$

$$i = 1, 2, \ldots, m$$

A^* consists of the most preferable desirability of every existing alternative in decision (Asgharpoor 2004). The various steps of this method will be presented as follows.

2.3.1 Making Dimensionless

To compare various measurement scales (for various attributes), it is necessary to use a dimensionless method (Asgharpoor 2004). There are several techniques to make dimensionless (none missing), but the normal method is explained. First normality of data is checked using the Shapiro-Wilk test. If data is not normal, Box–Cox technique can be used for normality. Finally, Uniform Function is applied to unify and perform dimensions of data.

In Box–Cox method, it is necessary to estimate a value for λ, and then the following equation can be applied for normality.

$$y_i = \frac{x_i^\lambda}{\lambda} \; for \; \lambda \neq 0 \tag{2.6}$$

where y_i = normalized data, x_i = original data and λ = a value which we substitute in Eq. (2.6).

2.3.2 Assessment of Weighting (Wj) for Attributes

In the majority of MCDM problems, particularly the part of MADM, it is essential to be acquainted with the relative significance of existing attributes, considering the sum of them should be equal to 1, and this relative significance assesses the preference rank of each attribute compared with other attributes for specified decision. There are various methods of weighting in MCDM. Considering the experts' opinions and the existing conditions, it is selected and weighted parameters that have a specific significance in the standard and have a high significance in special consumptions and also parameters which are given a higher worth for water quality attributes.

The final stage of decision is ranking of stations using Simple Additives weighting method (SAW).

In the SAW method vector w (weights of significant attributes that obtained in previous) is assumed and suitable alternative A* is calculated as follows:

$$A^* = \left\{ A_i \Big| \max \frac{\sum w_i r_{ij}}{\sum w_j} \right\} \quad (2.7)$$

And if $\sum w_j = 1$:

$$A^* = \left\{ A_i \,|\, \max \sum w_i r_{ij} \right\} \quad (2.8)$$

Final weights that were obtained in the previous part are used in this stage and multiplied to equalized and dimensionless values of MADM matrix, and are the calculated sum of the parameters in each line. Therefore, for each station, a number was obtained that can be our selection attribute and based on that the stations were specified.

2.4 Dynamic Programming Approach (DPA) Method

The DPA is one of most applicable technique which nowadays is applied for modeling of operation systems. The DPA is a method for solving the problems joint with the Multi-stage Decision. In this method concerning the characteristic of staging the problem, we solve the problem with n staging and single variable instead of solving the problem with *n* variables.

2.4.1 The DPA Theory

At the first stage, reduction of the network is achieved by dividing river catchment area to N ($K = 1, 2, ... N$) Primary basin. This division should not be according to the

2.4 Dynamic Programming Approach (DPA) Method

hydrology of the basin. The characteristic of the catchment area such as topography, geology, meteorology, land uses, population density and rivers crossing each other may be as criteria for dividing a catchment area to primary catchment area. Each sub catchment area must have a minimum one water quality monitoring station (Harmancioglu and Fistikoglu 1999).

In each primary catchment area named K, P_K is the number of existing stations in that catchment area, and R_k is the number of selected stations in the primary catchment area named K.

Thereby, the numbers of the possible cases of stations selection can be determined which if during determination of possible combination of stations which they must remain in each sub catchment area; in this case all the catchment area of the river must be considered. Number of substitution combinations can be obtained from binomial distribution (Eq. (2.9)).

$$C(TP_N, TR_N) = \binom{TP_N}{TR_N} = \frac{TP_N!}{TR_N!(TP_N - TR_N)!} \qquad (2.8)$$

where TP_N = the number of existing stations in all catchment area of the river, and TR_N = the number stations to be retained in all networks. When the catchment area is divided by N sub catchment area, K = 1,.... N, each primary catchment area has P_K of primary station which is part of total existing stations in all catchment areas (TP_N) (Eq. (2.9)) therefore:

$$TP_N = \sum_{K=1}^{N} P_K \qquad (2.9)$$

where TR_N = the sum of the number of existing stations in each sub-basin Eq. (2.10). When TR_N = total number of stations to be retained in the basin; number of stations which must remain in primary catchment area K = R_K, which can be defined according to Eq. (2.10).

$$TR_N = \sum_{K=1}^{N} R_K \qquad (2.10)$$

Consequently, number of feasible answers for each sub catchment area with availability of R_K can be determined by Eq. (2.11).

$$C(P_K, R_K) = \binom{P_K}{R_K} = \frac{P_K!}{P_K!(P_K - R_K)!} \qquad (2.11)$$

where R_K = 0, 1, 2, ..., P_K.

Therefore, the number of total combinations of catchment areas substitution in the K primary catchment area can be obtained by Eq. (2.12).

$$TASC_K = \sum_{R_K}^{P_K} C(P_K, R_K) = 2^{P_K} - 1 \qquad (2.12)$$

where $TASC_k$ = the number of total combinations of substitution catchment area in the primary catchment area. The R_K in each primary catchment area depends on R_K of other primary catchment area (Eqs. (2.2–2.10)). Total number of substitution stations in all catchment areas will be TASC (Harmancioglu and Fistikoglu 1999).

$$TASC = TASC_1 \times TASC_2 \times \cdots \times TASC_N \qquad (2.12)$$

Or

$$TASC = \prod_{K=1}^{N} TASC_K \qquad (2.12)$$

The TASC will assume a very large number according to the amount of $TASC_K$. For solving this problem, Letternmaier et al. (1984) suggested the stream order number method for the limitation of combination of substitution stations in the primary catchment area.

2.4.2 Normalization and Uniformization Procedure

For comparison of different scale measurements, data should be Dimensionless, thereby, the converted indices elements (n_{ij}) were computed with dimensionless quantity. Several methods are available to change the quantity to dimensionless. If the data was not normal, Box–Cox method was used for normality (Eq. 2.6). Subsequently for uniformization and dimensions the uniform function was used.

$SR_{j(i)kl}$ = the original data, $SU_{j(i)kl}$ = the normal and uniform data for the station i and sub catchment k.

For each quantity of TR_N, it is necessary to determine the number of selected stations in each primary catchment area k, named R_K. Therefore, the selected stations are the stations at which their sum of normalized data ($SU_{j(i)kl}$) are maximized.

We indicate $SU_{j(i)kl}$ with $TS_{j(i)K}$ (Eq. (2.14)).

$$TS_{j(i)k} = \sum_{l=1}^{l_N} SU_{j(i)kl} \qquad (2.14)$$

Where l_N = number of parameters in the station i and in sub catchment area k.

2.4 Dynamic Programming Approach (DPA) Method

While the significance of the parameters differs when they are compared with each other; we use relative weight for each parameter concerning the objective of monitoring, and the Eq. (2.14) will be converted to Eq. (2.15).

$$TS_{j(i)k} = \sum_{l=1}^{l_N} W_l \times SU_{j(i)kl} \qquad (2.15)$$

where, W_l = relative weight for parameter i.

By Eq. (2.15) can be obtained total(all) amounts of the parameters in primary catchment area k and in the station i. In each primary catchment area k, different selections of the stations which depends on R_k, and the amount of $TS_{j(i)k}$ in each station is different. Therefore, the calculations must be with the combinations which the amount of $TS_{j(i)k}$ will be maximized (Eq. 2.16).

$$MTS_{j(i)k} = \max TS_{j(i)k} \qquad (2.16)$$

By determination of the TR_N, the RK options are the selections which have a maximum amount of $MTS_{j(i)k}$ (Eq. (2.17)

$$SMTS = \max \sum_{K=1}^{N} \sum_{i=1}^{R_K} MTS_{j(i)k} \qquad (2.17)$$

Equation (2.17) has two dimensions for solving it, we applied the DPA.

The objective is to find the combination of the stations which the amount of $MTS_{j(i)k}$ will be maximized (Eq. 2.18). TR_N is counter with known amount.

The objective function is Eq. (2.18).

$$V = \max \sum_{K=1}^{N} \sum_{i=1}^{R_K} MTS_{j(i)k} \qquad (2.18)$$

The constraints are as follows:

$$\sum_{K=1}^{N} R_K = TR_N \qquad 0 \leq R_K \leq TR_N \qquad (2.19)$$
$$0 \leq j(i) \leq P_K, \qquad j(i) \neq j(h), i \neq h$$

where j(i) = the number of stations in primary catchment area K. V = the objective function, N = total number of primary catchment area, R_K = the number of stations which are retained in the primary catchment area, i = an index of the station in k primary station, j(i) = numbers of index stations i in the k primary station, and P_K = the number of existing stations in the k primary station.

2.5 Application of Sanders Method

In a study, designing the number of water quality monitoring stations on the Kārūn River, Iran, was carried out. We applied the Sanders method on the basis of irrigation and drinking water quality indices. The specified sites were compared to the existing sites in the system and the matched stations to this scheme were selected. Water quality sampling was carried out on 20 stations by the Iranian Department of Environment (DOE) in Khuzestan Province during the years 1995–2005. (Table 2.2 and Fig. 2.1).

At the first step, the catchment area (Fig. 2.2) was divided into four sub-basins which were named A, B, C and D. Gotvand Dam to Band-e-Gheer, Dez Dam to Band-e-Gheer, Band-e-Gheer to the south of Ahvaz City and Darkhoveyn to the Persian Gulf were considered as sub-basins A, B, C and D, respectively. Water quality parameters including the Biochemical Oxygen Demand (BOD), Chemical Oxygen Demand (COD) as degradable and Total Dissolved Solid (TDS) as the non degradable were selected which weighted by coefficients of 0.2, 0.3 and 0.5, respectively. The combined results of these coefficients were the basis of the evaluation and weighting of the amount of pollution discharged by each of the pollution sources. For this matter, the mass discharge of effluent BOD, COD and TDS along with their weighting percentage for each pollution source of all sub-basins were calculated (Tables 2.3, 2.4, 2.5 and 2.6). Then, by adding the number of entering branches, as presented in the Fig. 2.2, and allocation one of the number of the main branches of the river, the number of pollutants entering each pollution source was calculated.

As an example, for Pars Paper Factory in sub-basin B, BOD, COD and TDS weighting percentages due to their total values, are 19.87, 57.36 and 10.8. The pollution load of this factory is calculated as follows:

$$0.2(19.87) + 0.3(57.36) + 0.5(1.08) = 21.7 \approx 22$$

Table 2.2 Monitoring Station on the Kārūn River

Station number	Station name	Station number	Station name
1	Gotvand	11	Kārūn–Band-e-Gheer
2	Kārūn–Band-e-Mizan	12	Mollasani
3	Shotayt–Arab Asad	13	Zargan
4	Shotayt–Band-e-Gheer	14	Pol-e-panjom
5	Gargar–Shushtar	15	Ommotamir
6	Kargar–Band-e-Gheer	16	Darkhoveyn
7	Dez–Chamgolak	17	Saboon sazi
8	Dez–Ab-e-sheerin	18	Haffar
9	Dez–Mostoafi	19	Abolhasan
10	Dez–Band-e-Gheer	20	Choabadeh

2.5 Application of Sanders Method

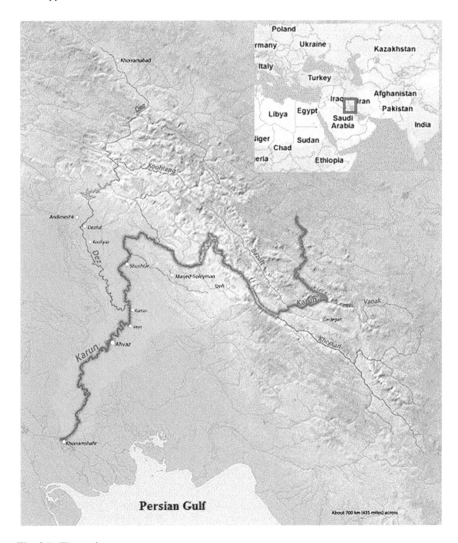

Fig. 2.1 The study area

Eventually, we determined the water quality of each station using the Sanders Method. The results described that the situation on stations located on Chobdeh, Abolhasan, Hafar, Soap Factory, Darkhoveyn, Omoltamir, Pol-e-Panjom, Zargan, Mollahasani and Kārūn-Band-e-Gheer was more crucial than others. On the other hand, the water quality of stations located on Gotvand and Dez-Chamgalk was more favorable. Totally the water quality of the Kārūn River for drinking and irrigation purposes in the sub-basins A and B is considerably better than sub-basins C and D (Table 2.7).

16   2 Selection of Water Quality Monitoring Stations

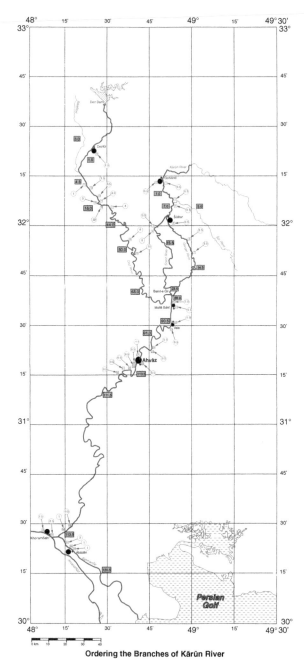

Fig. 2.2 Water quality sampling of the Kārūn River

2.5 Application of Sanders Method

Table 2.3 Amount of pollution load for sub-basin A

Pollution source location and number on the map	Weighting percentage BOD_5	Weighting percentage COD	Weighting percentage TDS	Final weighting
Kārūn sugar cane factory (2)	0.3	1	0.15	0.5
Fish cultivated field (7)	0.65	0.63	5.47	3.5
Kārūn sugar cane drainage (9)	3.4	7.3	10	8
Agheli drainage (8)	0.04	0.16	0.74	0.5
Janat Makan drainage GE (3)	0.015	0.01	0.31	0.5
Janat Makan drainage Gd (4)	0.02	0.02	0.21	0.5
Ghaghar agricultural drainage (6)	0.11636	0.5	5.98	3.5
Ghotvand (1)	0.596	0.21	0.15	0.5
Shoshtar (5)	1.097	0.57	0.19	0.5

Table 2.4 Amount of pollution load for sub-basin B

Pollution source location and number on the map	Weighting percentage BOD_5	Weighting percentage COD	Weighting percentage TDS	Final weighting
Dezful sugar factory (11)	2.9	0.63	0.25	1
Haft Tapeh sugar cane factory (15)	7.999	1.94	1.01	3
Pars paper production factory (16)	19.87	57.36	1.08	22
Saghari drainage (12)	0.4551	1.18	1.7	1.5
Solameh and agirob drainages (13)	1.21	3.86	3.82	3.5
Haft Tapeh drainage (14)	11.37	2.58	2.89	4.5
Meyanab drainage (18)	0.455	0.61	11.29	6
Kārūn sugar cane drainage (17)	0.96	0.84	6.67	4
Shoaybeh drainage (19)	0.2528	4.18	20.75	15

Table 2.5 Amount of pollution load for sub-basin C

Pollution source location and number on the map	Weighting percentage BOD_5	Weighting percentage COD	Weighting percentage TDS	Final weighting
Ramin power plant (23)	0.51	0.19	0.25	0.5
Zargan power plant (25)	0.081	0.05	0.19	0.5
Ahvaz sugar organization (24)	12.33	0.02	5.6	5.5
Sepanta factory (28)	0.045	0.01	0.06	0.5
Farsit factory (29)	0.0506	0.02	0.21	0.5
Ahvaz pipe production factory (30)	0.0976	0.02	0.07	0.5
National group factory (31)	0.6675	0.29	0.28	0.5
National group industrial plant (32)	0.1365	0.05	0.02	0.5
Ahvaz Khoramnoosh (26)	0.1766	0.18	0.3	0.5
Ahvaz slaughterhouse (20)	0.41	0.23	2.62	1.5
Mollahsani (22)	0.591	0.2	0.15	0.5
Vaes (22)	0.6118	0.38	0.12	0.5
Ahvaz city (27)	24.262	9.17	7.78	11.5

Table 2.6 Amount of pollution load for sub-basin D

Pollution source location and number on the map	Weighting percentage BOD_5	Weighting percentage COD	Weighting percentage TDS	Final weighting
Stris milk factory and Pasargad (33)	0.1264	0.05	0.34	0.5
Soap factory (34)	0.4955	1.4	0.24	1
Khorramshahr Khorram nosh soft drink production factory (35)	0.6068	0.05	0.04	0.5
Khorramshahr city (36)	2.4	0.97	1.26	1.5
Abadan refinery (39)	0.0	0.5	1.26	1.0
Abadan petrochemical production plant (37)	0.0	0.36	5.26	3.0
Abadan city (38)	1.03	0.5	0.84	1.0

2.5 Application of Sanders Method

Table 2.7 Score of water quality monitoring of the Kārūn River using Sanders method

Sub-basin	Location of station	Order
A	Gotvand	1
	Kārūn–Band Mizan	7
	Shotait–Arab asad	15.5
	Shotait–Band Gheer	15.5
	Ghargar–Shooshtar	7.5
	Ghargar–Band Gheer	14.5
B	Dez–Chamgalk	1
	Dez–Ab-e-sheerin	44
	Dez–Mostofei	50
	Dez–Band Gheer	65
C	Kārūn–Band Gher	88
	Mollahsani	90.5
	Zargan	91
	Pol-e-Panjom	110.5
	Omoltamir	111.5
D	Darkhoveyn	111.5
	Soap factory	113.5
	Hafar	115.5
	Abolhasan	118.5
	Chobdeh	120.5

2.5.1 Comment on the Application

Subjective selection of water quality monitoring may elevate a number of stations and may cause increasing errors in selection of the stations. As a result, the financial cost of installation and operation of the monitoring network can be increased. Therefore, mathematical methods which using experimental data such as Sanders method and dynamic programming approach may help to determine the locations and number of water quality monitoring stations, economically and correctly.

References

Asadollahfardi G, Asadi M, Nasrinasrabadi M, Fraji A (2014) Dynamic programming approach (DPA) for assessment of water quality network: a case study of the Sefīd-Rūd River. Water practice and technology 9(2):135–148

Asadollahfardi G, Khodadai A, Azimi A, Jafarnejad M, Shahoruzi M (2011) Multiple Criteria Assessment of Water Quality Monitoring System in Karoon River. J Int Environ Appl Sci 6 (3):434–442

Asgharpoor M (2004) Multi Criteria Decisions. Tehran university publication, Tehran

Bellman R (1957) Dynamic programming. Princeton University Press, Princeton

Cetinkaya CP, Harmancioglu NB (2012) Assessment of water quality sampling sites by a dynamic programming approach. J Hydrol Eng 17(2):305–317

Dandy GC (1976) Design of water quality sampling systems for network. Ph.D. Thesis, Massachusetts Institute of Technology, USA

Harmancioglu NB, Fistikoglu O (1999) Water Quality Monitoring Network Design. Kluwer Academic Publishers, Dordrecht, Netherland

Harmancioglu N, Alpaslan N, Alkan A, Ozkul S, Mazlum S, Fistikoglu O (1994) Design and evaluation of water quality monitoring networks for environmental management. Report prepared for the research project granted by TUBITAK, Scientific and Technical Council of Turkey, Project Code: DEBAG-23, Izmir (In Turkish)

Harmancioglu N, Cetinkaya C, Geerders P (2004) Transfer of information among water quality monitoring sites: assessment by an optimization method. In: 18th International conference informatics for environmental protection

Harmancioglu NB, Icaga Y, Gul A (2004) The use of an optimization method in assessment of water quality sampling sites. Eur Water 5(6):25–35

Harmancioglu NB, Alpaslan N (1992) Water quality monitoring network design: a problem of multi-objective decision making1. JAWRA J Am Water Resour Assoc 28(1):179–192

Hudak PF, Loaiciga HA (1993) An optimization method for monitoring network design in multilayered groundwater flow systems. Water Resour Res 29(8):2835–2845

Icaga Y (2005) Genetic algorithm usage in water quality monitoring networks optimization in Gediz (Turkey) river basin. Environ Monit Assess 108(1–3):261–277

Islamic Republic of Iran Meteorological Organization (1983) Climate Statistic http://www.irimo.ir. Accessed 15 Aug 2012

Karamouz M, Kerachian R, Akhbari M, Hafez B (2009) Design of river water quality monitoring networks: a case study. Environ Model Assess 14(6):705–714

Karemi M (2002) Design of river water quality monitoring system. MSc Theses, Amirkabir University, Iran (In Persian)

Khalil B, Ouarda T (2009) Statistical approaches used to assess and redesign surface water-quality-monitoring networks. J Environ Monit 11(11):1915–1929

Khalil B, Ouarda T, St-Hilaire A (2011) A statistical approach for the assessment and redesign of the Nile Delta drainage system water-quality-monitoring locations. J Environ Monit 13 (8):2190–2205

Letternmaier DP, Anderson DE, Brenner RN (1984) Consolidation of a stream quality monitoring network1. JAWRA J Am Water Resour Assoc 20(4):473–481

MacKenzie M, Palmer RN, Millard ST (1987) Analysis statistical monitoring network design. J Water Resour Plann Manage 113(5):599–615

Noori R, Kerachian R, Khodadai Drban A, Sakebaneya A (2007) Assessment of significance of monitoring stations rivers using analyses principal component and factor analys: a case study Kārūn River. J Wastewater, 18(63):60–69

Ozkul S, Centinkaya CP, Harmancioglu NB (2003) Design of water quality network in Gediz Basin using drainage and effluent discharge characteristic. In: Proceedings 1st national water engineering symposium. State Hydraulic Works (DSI) Publications, Ankara, Turkey, pp 215–227 (In Turkish)

Sanders TG, Ward RC, Loftis JC, Steele TD, Andrain DD, Yevjevich V (1983) Design of network for monitoring water quality. Water Resources Publication, Littleton, Colorado

Schilperoort T, Groot S (1983) Design and optimization of water quality monitoring networks. Delft Hydraulics Laboratory Delft, The Netherlands

Sharp W (1970) Stream order as a measure of sample source uncertainty. Water Resour Res 6 (3):919–926

Sharp W (1971) A topologically optimum water-sampling plan for rivers and streams. Water Resour Res 7(6):1641–1646

Ward RC, Loftis JC (1986) Establishing statistical design criteria for water quality monitoring systems: review and synthesis1. J Am Water Resour Assoc 22(5):759–767

Woldt W, Bogardi I (1992) Ground water monitoring network design using multiple criteria decision making ant) geostatistics1. J Am Water Resour Assoc 28(1):45–62

Chapter 3
Water Quality Indices (WQI)

Abstract Having a lot of data for different water quality parameters in surface water without proper interpretation are not useful for water quality management. Due to the extent of water quality parameters, water quality indices (WQI) could be used as a point scale for interpretation of these parameters. WQI is the essential prerequisite of water quality management. Since 1978, much effort have been done to present techniques to summarize water quality data to a defined numeric digit which describes the degree of water quality. In this chapter, at the first step, the historical background of WQI was reviewed. Afterward, National Sanitation Foundation's Water Quality Index (NSFWQI) method and British Colombia water quality index (BCWQI) Method that are used frequently, described in details. Finally, the application of NSFWQI in the Kārūn River and Sefīd-Rūd River, which located on the south-west and north of Iran, were described. In addition the WQI of the Sefīd-Rūd River was investigated by BCWQI.

3.1 Historical Background

Huge amount of water quality data without precise interpretation cannot help water quality management properly. Therefore, it is necessary to summarize water quality data to a defined numeric digit which indicates the degree of water quality. To solve this problem there is a technique which is called water quality indices. Water quality indices are methods that can help water quality management to determine existing water quality.

A survey on the types and extent to which water quality indices are being used in the USA was conducted by Ott (1978). It is; however, noteworthy that comparison of different indices is rather controversial since the underlying assumptions and the aim of the different applications vary (Canter 1985).

Schaefer and Janardan (1979) studied five water quality indices among which, two indices, P1 and I, employed to rank data and followed the beta distribution. These two indices, which are true statistical measures of water quality, can be used with any set of water quality parameters, and also correlate highly with biological and subjective engineering assessments of water quality. Schaefer and Janardan (1979) also described another three indices using raw material and Chi-square distribution. Two of these three mentioned indices, B1 and B2, do not exhibit sufficient bias to serve as a general measure of water quality, although they may prove very useful in some specific situations. The third index, index C, can be applied quantitatively on monitoring stations. Steinhart et al. (1982) presented an index to help summarize technical information on the status of, and trends in lake water quality. Although the index was developed for the near-shore water of the Great Lakes of North America, they claimed that the index concept was practicable to other temperate lakes of generally high water quality. They attempted to minimize the vulnerability of the index by employing the following strategies:

- Selecting variables relevant to the uses of the lake under study
- Constructing rating curves based on established criteria
- Focusing on variables for which data are available and reliable
- Providing a reporting format that supplements the overall index number
- Creating an index that can be revised easily to reflect new knowledge or changing priorities
- Noting possible pitfall in suitable places, and
- Providing an optional user guide on request to further safeguard against possible nuisances.

House (1990) compared the results of the WQI applied to data collected by the Severn Trent Water Authority for fiscal years 1978/1979 and 1979/1980, and the value of the information derived using this index to that produced by the application of the National Water Council classification employed by the United Kingdom water industry. The study highlighted a number of advantages of using an index over NWC classification. Smith (1990) studied the WQI in New Zealand. The index development in the country was linked to the legislation of recommended Water Quality Standards. The indices were intended to assist in the dissemination of water quality information, particularly to lay-people.

Barbiroli et al. (1992) proposed a new three structured method to obtain synthetic quality indices for air and water. The index is unique in the sense that besides the final indices, several intermediate indices are also computed, allowing the environmental managers to have indices at different degrees of aggregate. This particular methodology is permitted for:

- Selection of physical, chemical, and biological parameters for air and water, definition of the variability of the selected parameters, the transformation of the value assumed by the various parameters in sub indices characterizes by a variable interval from 0, which represents minimum quality to 10, the maximum quality;
- Eventual weighting of parameters; and
- Construction of intermediate indices and the final synthetic index through the use of a suitable aggregate function. The method was designed to be as general and objective as possible, while still having the luxury of being adaptable to fit particular situations.

3.1 Historical Background

Doylido et al. (1994) studied water quality index in the Vistula river basin in Poland. Bocci et al. (1994) investigated Marine bacteria as indicators of water quality. They also developed a water quality index during the course of their research. Erondu et al. (1993) studied the classification of the new Calaba River at Aluu Port Harcourt in Nigeria using an experimental model. The primary indicators of water quality were BOD, Dissolved Oxygen (DO), temperature, sulfide and other physic-chemical and biological parameters. Jonnalagadda and Mhere (2001) studied the water quality of Odzi River, the main river issuing forth from the Eastern Highlands of Zimbabwe, using water quality indices. They monitored chemical parameters namely temperature, conductivity, total suspended solid, BOD, total phosphate and nitrate for 6 months sampling stations during 9 months. The results indicated that while the water was medium to good quality in the upper stream, the quality vitiated downstream, possibly due to the seepage from the abandoned mine dumps and discharges from farm land which infiltrated the river. Vollenweider et al. (1998) developed a new trophic index (TRIX) based on chlorophyll, oxygen saturation, mineral and Total Nitrogen (TN) and phosphorus, which is applicable to coastal marine water. The index is scaled from 0 to 10 covering a wide range of trophic conditions from oligotrophy to eutrophy. Secchi disk transparency combined with chlorophyll; instead, define a turbidity index (TRBIX) that serves as a complementary water quality index. The two indices are combined in a General Water Quality Index (GWQI). Ladson et al. (1999) developed an index of stream conditions (ISC) to assist water management by providing an integrated measure of their environmental condition. The ISC scores five aspects of the stream conditions:

- Hydrology: evaluated by tabulating changes in volume and seasonality of the flow from the natural condition
- Physical form: assessed by the bank stability, bed erosion or aggregations, influence of artificial barriers, as well as the abundance and the origin of the coarse woody debris
- Stream side zone: based on the types of plants, spatial extent, width and intactness of the riparian vegetation, regeneration of over story species, and the condition of wetlands and billabongs
- Water quality: based on an assessment of phosphors, turbidity, electrical conductivity and pH
- Aquatic life: appraised by the number of macro invertebrate families present.

Karydis and Tsritisis (1996) assessed the efficiency of 12 ecological indices expressing diversity, abundance, evenness, dominance and the biomass of phytoplankton to describe the tropic levels in the coastal area in the Eastern Mediterranean. They found that some of the commonly used indices such as the Simpsons' Shannon and Mayalef indices did not perform quite satisfactorily when used to establish eutrophic trends. On the other hand, the Menhinick's index, Kothe 's index, Species Evenness, Species Number and Total Number of individuals proved effective for distinction among oligotrophic, mesotrophic and eutrophic waters. The most efficient among this lot; however, was the Kothe sand species number indices. Bordalo et al. (2001) studied the water quality of the Bangpakong River, the most important river basin in Eastern Thailand, using Scottish water quality indices. They collected samples from June 1998 through 1999 at 11 monitoring stations covering a total of 227 km of river path. The monitored parameters included temperature, DO, turbidity, suspended solid, pH, ammonia, fecal coli form, BOD, COD, phosphate, conductivity and heavy metals. The average WQI was found to be very low at a meager 41 % and quality of the water declined further during dry season. Schultz (2001) critiqued the index watershed indicator on the basis of these tests as well as some other considerations, and suggested that if the index were to be explicitly based on multi-attributed utility theory and methods, some of the difficulties could be considerably resolved.

Khan et al. (2003) studied the water quality of the three selected catchment areas of the Atlantic region: Mersey River, the Point Wolfe River and the Dunk River, using the Canadian Water Quality Index and British Columbia Water Quality Index. They also applied linear and quadratic models to analyze the water quality trends. The results demonstrated that the water quality trend for raw water (before treatment) used for drinking has improved considerably on the Point Wolf River.

Said et al. (2004) described the limitations of the application of a few water quality indices such as NSWQI, British Columbia Water Quality Index (BCWQI), Oregon WQI, Florida WQI and Watershed Enhancement Program WQI, and then they developed a new WQI which proceeds in two steps. The first step ranks the water quality parameters according to their significance. These parameters include DO, total phosphate, fecal coli form, turbidity, and specific conductivity. They give higher rank for DO than fecal coli form and total phosphate. Turbidity and specific conductivity, on the other hand, are given less influence. Jafarnejad (2005) studied water quality index of Kaurn River and using NSWQI. Debles et al. (2005) employed water quality index from nine physicochemical parameters, periodically measured through January 2000 to November 2000 at sampling stations on the Chillan River in central Chile. Their results showed that the river boasted good water quality in the upper and middle parts of the catchment area, but downstream, especially during dry season, the quality deteriorated. This was due to the discharge of urban wastewater into the river. They also applied principal component to modify existing water quality. The study indicated that the application of modified WQI reduces the cost associated with its implementation. Taebi et al. (2005) applied three eutrophication indices to define water quality in north east of Persian Gulf.

Bordalo et al. (2006) applied a modified nine–parameter Scottish WQI to assess the monthly water quality of the Doura River, an internationally shared River basin, during a 10 year period (1992–2001). The 98,000 km^2 of the Dorian River split between upstream Spain (80 %) and Portugal downstream (20 %). The water received by Portugal from Spain was of much reduced quality (WQI 47.3 ± 0.7 %). Quality; however, increased steadily downstream up to 61.7 ± 0.7. In general, the water quality in all of the monitoring stations was at best mediocre and often poor.

Lumb et al. (2006) applied Canadian Council of Ministers of the Environment Water Quality Index (CCMEWQI) to determine the water quality of the ackenzie-Great Bear Sub-basin. The results of their study showed that the water quality of the basin is impacted by high turbidity and total trace metals (mostly particulate) due to highly suspended sediment load during the open water season. Fernandez et al. (2004) reviewed 36 WQI and water pollution index (WPI); the results of their works showed that appreciable differences exist between different WQI on the same water sample. They concluded that the WPI developed in Colombia by Rmiirez et al. (1997) and the AMOEBA strategy which was developed by De Zwart (1995) in Netherland offered considerable advantages over more traditional formulations. Using the systems of Lake Kinneret, Israel and the Northern Lakes of Belarus, Parparov et al. (2006) demonstrated that water quality can be quantified to be part of sustainable management in relation to lake management activities. They used WQI and rating curve to reach their objective. For both Lake Kinneret and Naroch, they established rating curves under the assumptions that the conservation of the lake ecosystem is the prime objective of the resource managers. They proposed three separate levels of WQI integration, which are as follows:

- An expanded WQI system which, being the base system, was proper for describing different aspects of water resource uses. It served as a "common language" for communication between associates in lake management.
- Reduced system of WQI
- Composite water quality index (CWQI): This highest level of integration of water quality

Sedeno-Dias and Lopez-Lopez (2007) studied the water quality of the Rio Lerma catchment area, the notoriously polluted area of Mexico. They used water quality index multiplicative and weighted water quality index and principal component in their work. WQI scores judged the water unsuitable for drinking and demonstrated that it was vitally important to treat the water. Pinto et al. (2009) mainly reviewed and evaluated benthic community based biotic indices. They supplied a general overview of some indices premises and assumptions as well as their main advantages and disadvantages. Asadollahfardi (2009) applied NSFWQI method to surface water quality in Tehran and his results was satisfactory. Juttner et al. (2010) studied periphytic diatoms to evaluate the water quality of a newly created lake formed by the enclosure of the formerly tidal Cardiff Bay (Wales, UK) and the effects of two inflowing rivers which drain the densely populated and industrialized basin. They selected seven sampling stations in Cardiff Bay and two stations on the inflowing river and collected samples for diatoms and chemistry of water for 2 years. They used a revised UK trophic diatom index (TDI) and a new technique to figure out ecological quality ratios and ecological states classes as required by the EU Water Framework Directive. In the bay, diatoms reflected differences in river quality and possibly of local pollution in certain areas of the lake. The high rates of TDI indicate eutrophic to hypertrophic conditions in both the rivers and the bay and diatoms indicate poor ecological status.

3.2 Summary of WQI Methods

WQI can generally be divided into five groups, which are as follows:

- General WQI methods:
  These can be applied to determine the overall water quality. They do not; however, take into account the water usage.
- Specific WQI techniques:
  These are employed to define water quality for specific usage such as drinking, irrigation, industrial and protection of the aquatic life.
- Design Index:
  They enjoy immense use in decision making processes in management. In such cases, water quality classification is not taken into account. These methods are merely tools to help assess the impact of decisions and plan future measure for water management.
- Statistical Index:
  Statistical indices are perhaps the most objective methods in service for water quality classification. Since they use statistical models, little individual judgment is present.
- Biological Index:
  These classifications define water quality according to the effects of water on the aquatic life.

In each of the above mentioned general categories, researchers have developed many methods to assess water quality that employ diverse water quality parameters and mathematical equations.

Two methods of water quality indices described in detail which are as follows.

3.3 National Sanitation Foundation's Water Quality Index (NSFWQI) Method

With the support of the National Sanitation, Brown et al. (1970) presented an index for water quality. Their work can be summarized as follows:

First, Brown et al. (2000) prepared a questionnaire (No. 1) and sent it to a carefully chosen panel, whose members came from a variety of backgrounds including regulatory offices, local utilities management, consulting engineers, academics and waste control engineers. They were asked to consider 35 water quality parameters for possible selection in a water quality index. The panel members were also asked to give an importance rating for each parameter on a scale of 1–5, where 1 corresponded to the highest significance while 5 corresponded to the lowest significance.

To obtain greater convergence of opinion regarding the significance of each parameter for the index, a second questionnaire was prepared in which each member was asked to review the original rating while considering their peers'

3.3 National Sanitation Foundation's Water Quality Index (NSFWQI) Method

opinions and modify their choices, if desirable. However, little changes were noted in the modified responses.

Finally, the parameters which had emerged from the second questionnaire as being the most significant were presented that curves assigning variation in water quality for different water quality parameter values should be drawn for each parameter were requested from the panel. Two of these graphs are indicated in Figs. 3.1 and 3.2. In each figure, the solid line represents the arithmetic mean of the all of the respondents' curves, while the dotted lines bounding the shaded area

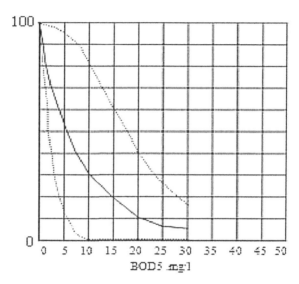

Fig. 3.1 Functional relationship for dissolved oxygen

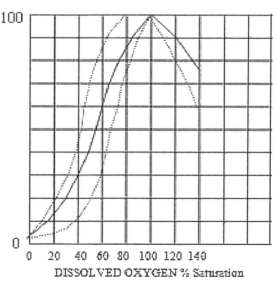

Fig. 3.2 Functional relationship for BOD

represent the 80 % confidence limits. Other rating curves for other parameters can be found in Ott (1978). The participants were also requested to give a relative significance grading on a five point scale to each parameter. This enables weighting for the parameters to be obtained.

The investigators sought to derive a set of weights for the index which would sum to 1.0, but more importantly, reflect the significance rating of the parameters by the panelist. The following three step procedure was used:

- The arithmetic means of significant rating were calculated from the parameters
- Temporary Weight was then calculated by the following formula:
  Temporary Weight = Significant rating of each parameter/Highest significant rating
- Final weights as sub index weights were calculated by:
  Final Weight = Temporary Weight/Sum of Temporary weights

The NSFWQI is calculated based on arithmetic and geometric means as given below:

$$NFWQI = \sum_{i=1}^{i=n} W_i I_i$$
$$NFWQI = \prod_{i=1}^{W} I_i \quad (3.1)$$

To calculate the index, the sub-index value I is read from the appropriate rating curves for pollutant parameter I, and then is multiplied by the sub-index weights calculated for each parameter, and summed over all the parameters. Tables 3.1 and 3.2 indicate the river classification using this method.

Table 3.1 Stream classification system suggested by NSFWQI (Ott 1978)

Representing color spectrum	Numerical range	Class
Red	0–25	Very bad
Orange	26–50	Bad
Yellow	51–70	Medium
Green	71–90	Good
Blue	91–100	Excellent
Red	0–25	Very bad

Table 3.2 Subdivisions of the WQI scale (House and Ellis 1980)

NWC Class	WQI range
1A	91–100
1B	71–90
2	41–70
3	21–40
4	10–20

3.3 National Sanitation Foundation's Water Quality Index (NSFWQI) Method

The NSFWQI is a subjective method and opinions of practicing professionals have affected the result of classification because the rating curves and intermediate weighting, the two parts of this technique, arise from experts' opinions. This method cannot describe temporary variation of surface water quality. On the other hand, this is a simple and easy method for analyzing water quality. All the parameters of water quality, which are employed in this method, are usually available in most monitoring programs. Most non technical people can Conveniently understand the result of classification. The NSFWQI relates to some other water quality index methods such as Horton, Prati and Dinius (Ott 1978) having a less ambiguous region for aggregation of sub-indices.

3.4 British Colombia Water Quality Index as (BCWQI) Method

The BCWQI method as an additive index in 1999 by the ministry of environment, land and parks of Canada was created to assess water quality. In this method, water quality parameters are calibrated with a certain limit and the amount of exceeding was determined. This limit can include recommended guidelines for maintaining the operational capability of the water in a certain design or each standard in which amount of the different uses of the water is. Therefore, so the use of standards in every area, region or country is one of the advantages of this method and the quality classification based on all measured parameters existed in each standard is possible. Equation (3.2) is used to calculate the final index.

$$BCWQI = \left[\sqrt{(F_1^2 + F_2^2 + (\frac{F_3}{3})^2}\right]/1.435 \qquad (3.2)$$

where F_1 = the percentage of parameters that have exceeded a certain limit, F_2 = the number of times generally that exceed certain limits, as a percentage of the total number of impressions and F_3 = the maximum exceeding of a certain limit (standard limit). The percentages of exceeding are defined as follows:

Percentage of exceeding = [(maximum allowed limit − measured value) /measured value]*100

Number 1.453 is chosen to ensure maximum number of BCWQI method reaches to number 100. Sampling frequency and increase the number of stations are the important notes that enhance the accuracy. About the disadvantage of this method it can be stated that this index is not able to describe the trend of water quality until are not exceeding from standard limits. Also, due to using the maximum exceed (F3), not specifying how many impressions are located above the maximum standard. Table 3.3 presents the interpretation of the pollution based on the BCWQI method.

In a specific use, high values of a parameter may be desirable, while in the other use this amount is not acceptable. Accordingly, some experts believe the

Table 3.3 Interpretation of the pollution based on the BCWQI method

Definition	Numerical value of the index
Excellent	0–3
Good	4–17
Suitable	18–43
Medium	44–59
Poor	60–100

classification should be conducted based on the type of the water use. Determining the specifications of the water quality will show how desirable for intended use. Hence, the water quality classification is the most important step in the management of water quality. Among the water quality indices, N.S.F. WQI is one of the most used indices and among the specially used indices (drinking purposes and agricultural), BCWQI is a newer and a more acceptable index. Therefore, application of the two methods is described.

3.5 Application of NFSWQI Method, a Case Study: Kārūn River (Iran)

The Kārūn River basin occupies an area of about 67,000 km^2, and is situated in the Khuzestan province in south of Iran. The excessive waste water discharge coupled with the withdrawal of pure water for domestic uses adds to the pollution in Kārūn River and has critically endangered aquatic life in the river. Since, it is one of the main rivers in Iran, water pollution in the Kārūn River system can significantly affect the development of Khuzestan Province and consequently, the economic development of the country since the province has a high potential for agricultural and industrial development. Therefore, maintaining the water quality of Kārūn River is of strategic significance. A large amount of used agricultural water returns to the rivers through drainage and return flows. However, because of its excursion through agricultural land, it returns with a high concentration of fertilizers, heavy metals, suspended and dissolved solids and pesticides. Therefore, it violates the national effluent standards. Agricultural and agro- industrial return flows, domestic wastewater of the cities and villages and industrial effluents are the main pollution point sources of the Kārūn River. The river system supplies the water demands of 16 cities, several villages, thousands of hectares of agricultural lands, and several hydropower plants. Ever increasing population, resulting in a hike in domestic water demands as well as industrial demand including but not limited to development of agricultural networks, fish hatchery projects, and inter-basin water transfers, darken the outlook for water quality of the Kārūn River Karamouz (2008) Fig. 3.3 indicates the water quality monitoring locations on the river.

In this case study, nine water quality parameters, DO, BOD$_5$, NO^3-, PO^4-, Total Solid (TS), pH, T, Turbidity and fecal coli form, was selected. To obtain the final weighting factor, 50 questionnaires were sent to faculty members of some of the

3.5 Application of NFSWQI Method, a Case Study: Kārūn River (Iran)

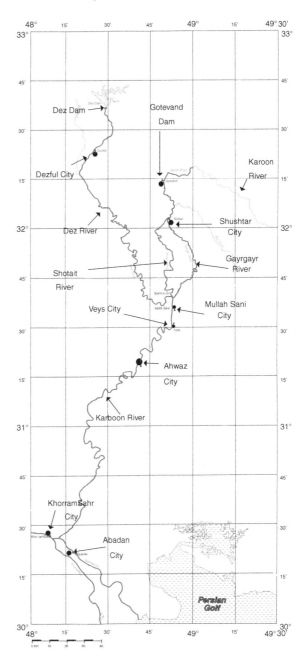

Fig. 3.3 Map of the Kārūn River

universities in Iran as well as water quality experts and managers in Water and Waste Water Company in Tehran. Only eighteen people were able to respond. Then, according to the mentioned methodology, final weighting factor was calculated (Table 3.4). If final weighting of the study and NSFWQI are compared, the results

Table 3.4 The results of opinion for Kārūn River Water Quality data (Asadolahfardi et al. 2005)

Water quality parameters	Mean of all significance	Temporary weights	Final weight
DO	1.89	0.83	0.13
BOD_5	1.56	1.00	0.15
pH	2.33	0.67	0.11
NO_3	2.11	0.74	0.11
PO_4	2.72	0.57	0.09
TS	2.33	0.67	0.1
Temperature	3.44	0.45	0.07
Turbidity	2.40	0.65	0.1
Fecal coli form	1.67	0.93	0.14

Table 3.5 Comparison of NSFWQI weighting factors and weighting factor in this case study

Water quality parameters	NSFWQI weighting factors	Weighting factor in this study
DO	0.17	0.13
BOD_5	0.15	0.11
pH	0.10	0.11
NO_3	0.11	0.10
PO_4	0.09	0.10
TS	0.10	0.07
Temperature	0.07	0.10
Fecal coli form	0.14	0.15

show that the two final weightings are about the same, and the maximum differences belong to BOD and temperature (Table 3.5). Therefore, the use of the questionnaire responses may be acceptable to apply to the study. Finally, water quality was computed according to mentioned methodology. Table 3.6 presents the results.

One of the advantages of the NSFWQI is its simplicity and availability of water quality parameters in the most water quality monitoring programs, and again, as previously mentioned, while the subjectivity of the method's is a disadvantage, it caused minimum effects in the final result. Another notable disadvantage is the method inability to indicate water quality trends, but that was not among the objectives of this case study. The aim of this case study was to clarify the existing situation of water quality and show the application of the methodology.

As described in Table 3.6, Kārūn River water quality from Shahhid Abahpoor to Shoshtar city was average, but as water circulated through the city, in Gargar branch, water quality worsened due to the discharge of raw domestic wastewater to the river. Dez River from Dez dam to Band Gheer and from Band Gheer to Ramin water station also possesses medium water quality. The water quality of the Dez River in Ahvaz city of Zargan monitoring station is not satisfactory, since industrial and domestic wastewater discharges to the river at that point. The quality of the water moving toward the Darkhovain monitoring station was middling as well.

3.5 Application of NFSWQI Method, a Case Study: Kārūn River (Iran)

Table 3.6 Final NSFWQI and interpretation of water quality of Kārūn River

Monitoring station	Final WQI	Water quality interpretation	Monitoring station	Final WQI	Water quality interpretation
Dez-Dez dam	59.42	Medium	Kārūn-Bandegher	51.69	Medium
Dez-Chamgalak	66.38	Medium	Kārūn-Ramin	51.76	Medium
Dez-Sugar factory	60.96	Medium	Kārūn-Zargan	49.04	Bad-medium
Dez-Abshrin	56.64	Medium	Kārūn-Newsite	51.45	Medium
Dez-Mostofei	53.29	Medium	Kārūn-Polpanjom	50.77	Bad-medium
Dez-Bandegher	58.10	Medium	Kārūn-Choneibeh	49.15	Bad-medium
Kārūn-Abbaspour dam	65.74	Medium	Kārūn-Omotayer	49.45	Bad
Kārūn-Gatvand dam	61.96	Medium	Kārūn-Darkhovein	51.75	Medium
Kroon-Bandmezan	61.08	Medium	Kārūn-Nahr	51.16	Medium
Gargar-Bandshushtar	50.11	Medium	Kārūn-Saboonsazi	47.15	Bad
Shotait-Shushtar	51.72	Medium	Hafar-Gomrok	58.58	Medium
Gargar-Bandegher	51.62	Medium	Bahmansher-Abolhasan	59.05	Medium
Shotait—Bandegher	52.89	Medium	Bahmansher-Choedeh	58.08	Medium

Subsequently, around Khoramshahr city, water qualities aggravates. River water quality from this point to the nearby Persian Gulf again improves to medium. It is worth mentioning that Noroozian (2000) also studied the water quality of the river albeit using the fuzzy method and obtained different results. The explanation lies in the fact that the NSFWQI is not for a specific application, while in the fuzzy study, standards of protection of aquatic life was considered. In addition, the definition of water quality in NSFWQI and fuzzy method is not quite the same.

3.6 Application of NSFWQI Method in Sefid-Rud River (Iran)

The case study area is Sefid-Rud River which is located in the province of Gilan, Iran (Fig. 3.4). This river originates from branches of Ghezelozan and Shahrod to Caspian Sea. The branch of Ghezelozan originated from the mountains of Kurdistan

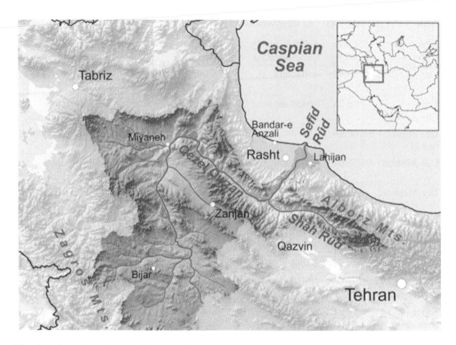

Fig. 3.4 Location of the Sefid-Rud River

and Azarbaijan with the maximum flow of 2,000 m³/s and the minimum of 4 m³/s. Another branch is Shahrud that is originated from the mountains of Alamot and Talaghan with the maximum flow of 800 m³/s and the minimum flow of 6 m³/s (Figs. 3.5 and 3.6). The dam constructed on the river is located at 200 km to northwest of Tehran and 100 km to Caspian sea at the confluence of two rivers of Ghezelozan and Shahrod. Basin's area is 57,880 km² and length of this river is 670 km; the basin is located in orbit of 49.9422° eastern and 37.4692° northern.

Water samples collected from upstream and downstream of the Sefid-Rud dam which it's location is illustrate in Fig. 3.5 for upstream and Fig. 3.6 for downstream.

The aim of this case study was to indicate the application of NSWQI methods for general water quality index and using BCWQI techniques for drinking purposes and agriculture in Sefid-Rud River.

The mean of the water quality data of the 21 stations which was monitored by Water Research Institute (Iran) between 2005 and 2006 is presented in Table 3.7. In this part, the numerical value of the NSFWQI method calculated for each of the 21 qualitative monitoring stations which is shown in Figs. 3.5 and 3.6. Equation 3.1 used to calculate the NSWQI. Due to lack of reliable data for fecal coli form and turbidity for all the stations, other parameters include DO, BOD_5, PO_4, NO_3, pH, TS and temperature were used to compute the WQI. Figure 3.7 indicates the results. As indicated in Fig. 3.7, according to the NSFWQI method the GSW5 station with the 47.41 had poor water quality and other stations were in the medium water quality. The ST2.1 station with 66.9 was in medium quality and had better condition than others.

3.6 Application of NSFWQI Method in Sefid-Rud River (Iran)

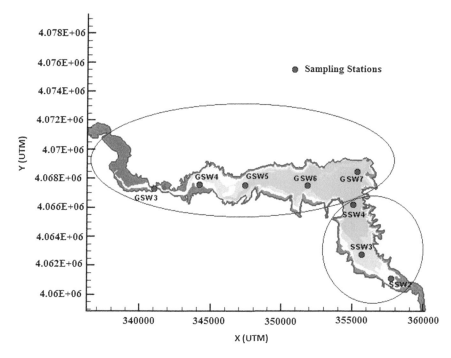

Fig. 3.5 The sampling stations of upstream

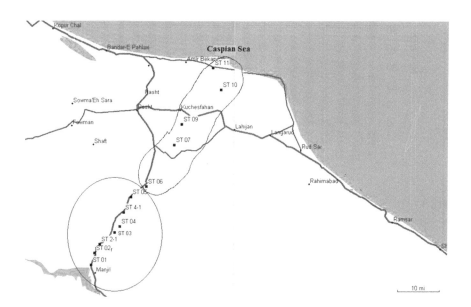

Fig. 3.6 The sampling stations of downstream

Table 3.7 The mean of water quality data for 21 stations in the case study area

Station	DO (mg/l)	Temperature (°C)	pH	TS (Mg/L)	PO_4 (Mg/L)	NO_3 (Mg/L)	COD (Mg/L)	BOD_5 (mg/l)
ST 01	5.6	17.9	8.2	43.9	0.5	4.1	16.5	8.2
ST 02	7.4	17.8	8.2	103.2	0.3	4.3	19.2	9.6
ST 2.1	9.5	18.1	8.2	86.9	0.2	4.3	22	11
ST 03	7.3	18.4	8.2	255.8	0.2	4.9	18.4	9.2
ST 04	6.4	19.1	8.3	375.3	0.1	4.4	16.3	8.2
ST 4.1	8	19.3	8.2	369.5	0.1	4.4	24.2	12.1
ST 05	8.9	18.9	8.2	143.8	0.1	4.6	26.4	13.2
ST 06	6.4	19.6	8.2	104.7	0.2	4.7	24.9	12.5
ST 07	7.6	21.8	8.1	156.0	0.2	3.2	30.2	15.1
ST 08	7.6	22.2	8.1	122.3	0.2	3.2	26.5	13.2
ST 09	7.9	21.6	8.1	101.1	0.1	3.6	25.1	12.6
ST 10	5.8	22.1	8.1	115.8	0.2	2.8	25.4	12.7
ST 11	6.1	22.2	8.1	130	0.2	2.7	26.8	13.4
GSW3	8.9	20	7.3	153.6	0.2	4.9	33	15.8
GSW4	9.5	17.6	8.0	37.2	0.2	4.6	25	12.6
GSW5	7.6	14.4	5.1	200.3	0.3	4.6	18	8.7

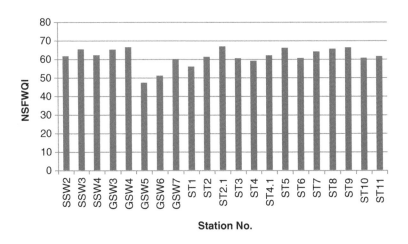

Fig. 3.7 The result of the NSFWQI of the Sefid-Rud River

3.7 Application of BCWQI Method in Sefid-Rud River (Iran)

The BCWQI method, which is an index for special purposes, was applied for drinking purposes and agriculture of the 21 stations on the Sefid-Rud River (Figs. 3.5 and 3.6). The parameters were used including DO, BOD_5, PO_4, NO_3, pH, TS and temperature.

3.7 Application of BCWQI Method in Sefid-Rud River (Iran)

Thus, the different monthly measured values were selected in each station (Table 3.7). Figures 3.8 and 3.9 show the results of the method based on drinking purposes and agriculture putting these parameter data into Eq. 3.2 Accordingly, when a numerical index is high the station is more critical, the ST9 station with the 104 index for the agriculture and the ST6 station with 2909 index for drinking purposes were at the critical situation. The SSW2 station with BCWQI method was 38 for the agriculture and 62 for drinking purposes. The two stations were the suitable water quality. Generally, the river's water for drinking purposes is not at the suitable situation; therefore, so, for drinking purposes need to be treated. Except for the ST9 station, the river's water for agriculture uses was in the media situation.

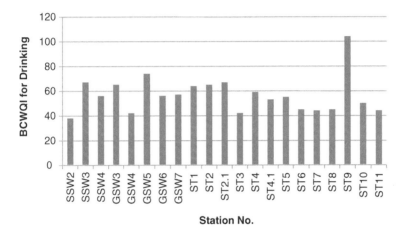

Fig. 3.8 The results of BCWQI method for of Sefid-Rud River regarding drinking uses

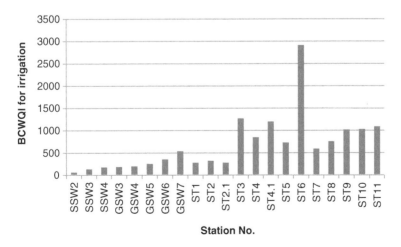

Fig. 3.9 The amounts of BCWQI of the Sefid Rud River for 2005–2007 data regarding irrigation uses

3.8 Comments on Application

The study describes:

The NSFWQI method has easy application and subjective technique.

According to the results of the NSFWQI method, the water quality of the Sepidrud River from the environmental view was medium station and the ST2.1 station had the best water quality and the GSW5 station was in the worst condition.

According to the results of the BCWQI method, the water quality of the Sepidrud River for drinking purposes was not suitable and needs to be fully treated. The ST6 and the SSW2 stations were the critics and suitable water quality, respectively.

Application of BCWQI method showed the river water quality for agriculture uses was in a medium situation. The ST9 and the SSW2 stations were in critical and suitable situation, respectively.

Considering the available data, the mentioned methods to define water quality may be acceptable tools.

References

Asadollahfardi G (2009) Application of water quality indices to define surface water quality in Tehran. Int J Water 5(1):51–69

Bacci E, Bucci M, Sbrilli G, Brilli L, Gambassi F, Gaggi C (1994) Marine bacteria as indicators of water quality. Chemosphere 28(6):1165–1170

Barbiroli G, Mazzaracchio P, Raggi A, Alliney S (1992) A proposal for a new method to develop synthetic quality indices for air and water. J Environ Manage 36(4):237–252

Brown RM, McLelland NI, Deininger RA, Tozer RG (2000) A water quality index Do We Dare? water and sewage works 339–343

Bordalo A, Nilsumranchit W, Chalermwat K (2001) Water quality and uses of the Bangpakong River (Eastern Thailand). Water Res 35(15):3635–3642

Bordalo AA, Teixeira R, Wiebe WJ (2006) A water quality index applied to an international shared river basin: the case of the Douro River. Environ Manage 38(6):910–920

Canter LW (1985) River water quality monitoring. University of Oklahoma, Norman

Debels P, Figueroa R, Urrutia R, Barra R, Niell X (2005) Evaluation of water quality in the Chillán River (Central Chile) using physicochemical parameters and a modified water quality index. Environ Monit Assess 110(1–3):301–322

Dojlido J, Raniszewski J, Woyciechowska J (1994) Water quality index–application for rivers in Vistula river basin in Poland. Water Sci Technol 30(10):57–64

Fernández N, Ramírez A, Solano F (2004) Physico-chemical water quality indices-a comparative review. BISTUA 2(1):19–30

Jafarnejad M (2005) Assessment of existing water quality monitoring stations and numbers of samples of the Kārūn River. MSc Thesis, Tehran University, Tehran

Jonnalagadda S, Mhere G (2001) Water quality of the Odzi River in the eastern highlands of Zimbabwe. Water Res 35(10):2371–2376

References

Jüttner I, Chimonides PJ, Ormerod SJ (2010) Using diatoms as quality indicators for a newly-formed urban lake and its catchment. Environ Monit Assess 162(1–4):47–65

House M (1990) Water quality indices as indicators of ecosystem change. Environ Monit Assess 15(3):255–263

Karydis M, Tsirtsis G (1996) Ecological indices: a biometric approach for assessing eutrophication levels in the marine environment. Sci Total Environ 186(3):209–219

Karamouz M, Kerachian R, Nikpanah A, Akhbari M (2008) Development of a MIS for river water quality data analysis 4(1):9–27

Khan F, Husain T, Lumb A (2003) Water quality evaluation and trend analysis in selected watersheds of the Atlantic region of Canada. Environ Monit Assess 88(1–3):221–248

Ladson AR, White LJ, Doolan JA, Finlayson BL, Hart BT, Lake PS, Tilleard JW (1999) Development and testing of an index of stream condition for waterway management in Australia. Freshw Biol 41(2):453–468

Lumb A, Halliwell D, Sharma T (2006) Application of CCME Water Quality Index to monitor water quality: a case study of the Mackenzie River basin, Canada. Environ Monit Assess 113 (1–3):411–429

Noroozian K, Tajrishy M, Abrishamachi A (2000) Water quality zoning of rivers by the technique of Fuzzy clustering analysis 20(1):20–30

Ott WR (1978) Environmental indices: theory and practice, Ann Arbor Science publication. INC, Ann Arbor Michigan

Parparov A, Hambright KD, Hakanson L, Ostapenia A (2006) Water quality quantification: basics and implementation. Hydrobiologia 560(1):227–237

Pinto R, Patrício J, Baeta A, Fath BD, Neto JM, Marques JC (2009) Review and evaluation of estuarine biotic indices to assess benthic condition. Ecol Ind 9(1):1–25

Said A, Stevens DK, Sehlke G (2004) An innovative index for evaluating water quality in streams. Environ Manage 34(3):406–414

Schaeffer D, Janardan K (1979) Development and applications of five new water quality indices. Bio J 21(6):539–552

Schultz M (2001) A critique of EPA's index of watershed indicators. J Environ Manage 62 (4):429–442

Sedeño-Díaz JE, López-López E (2007) Water quality in the Río Lerma, Mexico: an overview of the last quarter of the twentieth century. Water Resour Manage 21(10):1797–1812

Smith DG (1990) A better water quality indexing system for rivers and streams. Water Res 24 (10):1237–1244

Steinhart CE, Schierow LJ, Sonzogni WC (1982) An environmental quality index for the great Lakes1. JAWRA J Am Water Resour Assoc 18(6):1025–1031

Taebi S, Etemad-Shahidi A, Fardi GA (2005) Examination of three eutrophication indices to characterize water quality in the north east of Persian Gulf. J Coastal Res 405–411

Vollenweider R, Giovanardi F, Montanari G, Rinaldi A (1998) Characterization of the trophic conditions of marine coastal waters, with special reference to the NW Adriatic Sea: proposal for a trophic scale, turbidity and generalized water quality index. Environmetrics 9(3):329–357

Zwart D De, Trivedi RC (1995) Manual of water quality evaluation, RIMV, the Netherlands

Chapter 4
Time Series Modeling

Abstract When in surface water such as rivers, ponds and lakes detailed characteristic are not available, especially in developing countries, application of deterministic model are not applicable. In this regards, stochastic modeling are applied for estimating the future value of water quality parameters. There has been much effort in developing this technique for solving other engineering matters. Time series modeling as a stochastic model is trying to make probabilistic statements about the relation between system components and their future values and is used frequently in water quality management. In this chapter, after a preliminary explaining historical background of the method and introducing of time series modeling, the various methods such as Box-Jenkins methodology including stationary and non-stationary models and seasonal and non-seasonal models, exponential smoothing methods and Winter's method, was stated and described in detail. Finally the application of time series modeling on Latian Dam, which located in the southeastern part of Tehran province in Iran, was discussed.

4.1 Introduction

Planning a water pollution control program in rivers requires the following steps:

- The design of an experimental program and analyzing the water quality parameters,
- The study of mathematical methods for fitting equations to the parameters,
- Forecasting the future values of the parameters, and
- The development of appropriate control strategies.

Mathematical modeling plays a major part in formulating control strategies. Building mathematical models require special techniques, whereby one can express the natural phenomenon by mathematical equations. In general, there are two approaches to mathematical modeling, as stated below:

1. Deterministic, and
2. Stochastic.

The deterministic approach is based on a cause–effect relationship, and the models could occasionally be built by knowing the structural relations between components of a system. These relations are usually expressed in terms of differential equations whose solutions provide the desired models. Considering the independent variable is time, the resulting model can explain the variation of a system component or the whole system with respect to time. This makes it possible to forecast the future values of time dependent trends and other characteristics quite accurately. However, due to the presence of various unknown factors in nature, models are not completely deterministic. Therefore, in most cases it is required to study various phenomena under uncertain conditions, resulting from the effects of unknown factors. This leads to consideration of stochastic modeling.

In this approach, one is only able to make probabilistic statements about the relation between system components and their future values. A good example of this approach is stochastic hydrology. Numerous studies have proved the suitability of statistical methods in various hydrological problems. Especially, when the variation of the magnitude of one or more parameters of water quality is studied as a function of time, statistical method known as time series analysis are very helpful. By these methods, one can identify the trends, the periodic changes, and the random parts present in the natural series of data. Identification, estimation, and subsequent synthesis of these model components envisage the future path of the series, which in turn could help to take control measures (Bowerman and O'Connell 1987).

4.2 Historical Background

The emergence of stochastic hydrology goes back to 1914 when estimated the "probability of dry year". The arrival of digital computers provided suitable means for application of complex methods for statistical analysis such as time series.

Early applications of time series approach analyzing water resources were undertaken by Thomann (1967) who studied the time variation of temperature and dissolved oxygen of the Delaware Estuary. The data were obtained by continuously recording monitoring stations, operated jointly by the US Geological Survey Department and the city of Philadelphia. Carlson et al. (1970) and McMichael and Hunter (1972) have reported the successful use of Box-Jenkins method for time series analysis. The former applied this method to model and forecast annual stream flow data, where significant reductions in variance with one or two parameters is reported to be achieved. Other researchers use this method in developing models for daily temperature and flow in rivers. These models also incorporated deterministic components, which was preferable from a numerical and a rational point of view to a purely stochastic or purely deterministic model.

The Box-Jenkins method of time series analysis was applied in modelling the hourly water quality data recorded in the St. Clair River near Corunna, Ontario for chloride and dissolved oxygen level by Huck and Farquhar (1974). The models were parsimonious and physically reasonable and successful results were obtained.

Autoregressive and the first difference in moving average models represented the chloride data have been used to study the monthly water quality data in Chung Kang river located at the northern part of Miao-Li Country in the middle of Taiwan. Five years of investigation were conducted and 12 monthly water quality parameters were studied. The result was that forecasting with seasonable data seems to perform well when the Box-Jenkins technique is combined with non-parametric transformation.

Jayawardena and Lai (1989) undertook a very large research program. A time series analysis approach was applied to model the monthly COD values in the Yuancan and Fangcan and Guangzle reach of the Pearl River in southern mean China for a 21 year period. The basic properties of the water quality were determined, time and frequency-domain analyses were carried out, and the various stochastic models represent the dependent stochastic component. Using the probability distribution of the independent residuals generated synthetic water quality data, and the future water quality was forecasted.

For handling missing data and analyzing water quality data, Mahloch (1974) demonstrated the application of multivariate statistical techniques. The results of the study indicate that a simultaneous multiple regression technique may be used for supplying the missing observations and that at any particular level, the entire data matrix may be considered, thereby reducing the computational effort.

The sequential order of observations is a key concept, which is incorporated in stochastic modeling, especially in time series models. The 1960's witnessed a keen interest in the probabilistic structure of the sequence dependence of observations. It brought up the application of autocorrelation, an essential tool in the analysis of Autoregressive Integrated Moving Average (ARIMA) models.

To establish a working language, a concise description of time series analysis and its application in water quality studies is given in the following section.

Considering the deficit of water in Iran, protection of water resources against pollution is vital. In this regard, water quality monitoring is a tool which produces up to date information. Having a great amount of raw data without interpretation is not sufficient and it is necessary to analyze data and predict the variation of water quality in the future for any decision making on water quality management. Recently, more researchers have become interested in the application of time series models for the prediction of water quality.

Time series approach for analyzing water resources were first applied by Thomann (1967) who studied variation of temperature by the time and dissolved Delaware Estuary. The data was obtained by continuous recording by monitoring stations, operated jointly by the U.S. Geological Survey Department and the city of Philadelphia. Carlson et al. (1970) and McMichael and Hunter (1972) reported the successful use of the Box-Jenkins method for time series analysis.

The Box-Jenkins method for the time series analysis was applied to model the hourly water quality data recorded in the St. Clair River near Corunna, Ontario, for chloride and dissolved oxygen levels by Huck and Farquhar (1974), the models were physically reasonable and successful results were obtained. Autoregressive (AR) and first difference moving average (MA) models represented the chloride data well. Lohani and Wang (1987) also reported to have used this model to study the monthly water quality data in the Chung Kang River located at the northern part of Miao-Li County in the middle of Taiwan. Jayawardena and Lai (1991) applied an adaptive Auto Regressive Moving Average (ARMA) model approach for water quality forecasting. MacLeod and Whitfield (1996) analyzed water quality data using Box-Jenkins time series analysis of the Columbia River at Revelstoke. Caissie et al. (1998) studied water temperature in the Catamaran Brook stream. The short-term residual temperatures were modeled using different air to water relations, namely a multiple regression analysis, a second-order Marcov for process, and a Box-Jenkins time series model. Asadollahfardi (2002) applied Box Jenkins and Exponential smoothing models to monthly surface water quality data in Tehran for 3 years. Most of the models indicated seasonality. Kurunc et al. (2004) applied Auto Regressive Integrated Moving Average (ARIMA) and Thomas—firing techniques for 13 years of monthly data about the Duruacasu station at Yesilirmark River. Asadollahfardi et al. (2012) also worked about water quality of Jaj-Rud River and applied ARMA time series models.

4.3 Time Series

A time series is a chronological sequence of observations on a particular variable, such as daily, monthly or annual air or water quality, daily mean temperature, and so on (Box and Jenkins 1976).

Time series data are often examined in the hope of discovering a historical pattern that can be exploited in the preparation of a forecast. To identify this pattern, it is often convenient to think of a time series as consisting of several components. The components of a time series are trend, cycle, seasonal variations, and irregular fluctuations (Haan 1977).

Trend refers to the upward or downward movement, which characterize a time series over a period of time. Thus, the trend reflects the long-run growth or decline in time series. Its movements can represent a variety of factors. For example, long-run downward movements in DO might be due to gradual pollution (Bowerman and O'Connell 1987; Haan 1977).

Cycle refers to recurring up and down movements around trend levels. These fluctuations can have a duration of 2–10 years or even longer, measured from peak to peak or trough to trough. For instance, in arid zones, a cycle of an 11 year length is expected, for dry and wet years.

Seasonal variations are periodic patterns in a time series which repeat themselves in a certain length of time, (e.g. in a calendar year). Seasonal variations are usually caused by factors such as weather and customs (Cruz and Yevjevich 1972). The obvious example is the average monthly temperature, which is clearly seasonal in nature. Other examples are those of water quality parameter related to temperature. Average DO is high in winter and lower in the summer months, exhibiting a seasonal pattern.

4.3 Time Series

Irregular fluctuations are erratic movements in a time series, which follow no recognizable or regular pattern. Such movements represent what is "left over" in a time series after trend, cycle, and seasonal variations have been accounted for. Many irregular fluctuations in the time series are caused by "unusual" events that cannot be forecasted. Events such as earthquakes, accidents, wars, and strikes might cause irregular fluctuations. These can also result from errors made by the time series analyst.

The time series components already discussed above, do not always occur alone. They can occur in any combination or can occur altogether. Therefore, no single best forecasting model exists. A forecasting model suitable for forecasting a certain component alone may not suit the other components. Thus, one of the most important problems in forecasting is matching the appropriate forecasting model to the pattern of the available time series data (Box and Jenkins 1976). In Fig. 4.1 indicated time series exhibiting trend, seasonal and cyclical components.

There are many forecasting methods that could be used to predict future values of a time series. These methods can be divided into two broad categories-qualitative methods and quantitative methods (Bowerman and O'Connell 1987).

Qualitative forecasting methods, generally, use the opinion of experts to subjectively predict future events. These methods are often used when historical data are either not available at all, or are scarce. Qualitative forecasting techniques are also used to predict the changes in historical data patterns. Since the use of these data to predict future events are based on the assumption that the pattern of historical data will persist, changes in the data pattern cannot be predicted. The qualitative methods will not be used to analyze the current research data, so the discussion, on this method is ignored.

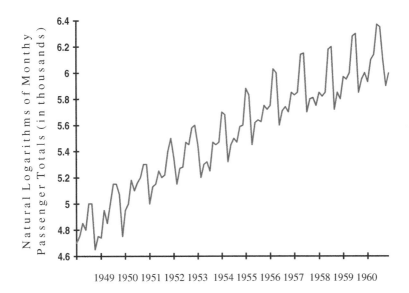

Fig. 4.1 Time series exhibiting trend, seasonal and cyclical components 4.3 forecasting methods

Quantitative methods of forecasting techniques, involve the analysis of historical data in an attempt to predict future values of a variable of interest. Quantitative forecasting models can be grouped into two kinds-univariate models and causal models.

Univariate models are the common types of quantitative forecasting methods. Such models predict future values of a time series, solely on the basis of the past values of the time series. When using such models, historical data are analyzed in an attempt to identify a data pattern. Then, if it will prevail in the future, this pattern is projected into the future to produce forecasts. Univariate forecasting is therefore; most useful when conditions are expected to remain the same. Changes, which are functionally related to time, can be incorporated into this method.

The use of causal forecasting involves the identification of other variables, which influence the variable to be predicted. Having identified these related variables, a statistical model such as time series regression or transfer function analysis is used to describe the relation between these variables and the variable to be forecasted. The statistical relationship derived is then used to forecast the variable of interest (Matalas 1966; Kisiel 1969).

For example, the BOD might be related to active bacteria and so on. In such a case, BOD is referred as the dependent variable, while the other variables are referred to as the independent variables. The analyst's task is to statistically estimate the functional relationship between the BOD and the independent variables. Having estimated this relationship with a good degree of confidence, it can be used to predict the future values of the independent variables to predict the future values of BOD (the dependent variable). These models are advantageous because they allow us to evaluate the impact of various alternate policies.

4.4 Forecast Error

Unfortunately, all forecasting situations involve some degree of uncertainty. This fact is recognized by including an irregular component in the description of a time series. This term, which is referred to as an error term, is the resultant of all errors made in the measurement, save for model choice. Given the model is correctly chosen, its parameters estimated from the data. Then this estimated model is used for prediction purposes. Thus, the forecast error for a particular forecast value is the difference between its actual value when it is observed and the predicted value obtained from the model. Let us denote an observation made at the time t by y_t and its forecast by $\hat{y}_t$. Thus, the forecast error is given by:

$$e_t = y_t - \hat{y}_t \tag{4.1}$$

Examination of forecast errors over time can often indicate whether the forecasting technique, being used, does or does not match the data pattern. For example, if a forecasting technique accurately forecasts the trend, seasonal, and cyclical components, which are present in a data pattern, the forecast error should reflect only the irregular component. That is, in such a case, forecast errors should appear purely randomly. Any sign of non-random nature present in forecast errors reveals the flaw in the model, which calls for the modification of the model. For example, if forecast errors describe a seasonal pattern, it is required to include a seasonal component in the initial model.

If the forecasting errors of the forecasting methodology over time is appropriate (i.e., errors are randomly distributed), it is important to measure the magnitude of errors in order to evaluate the accuracy of forecasts. As the "average errors" become larger, the forecasts get poorer. To this end, a measure often used is defined as:

$$MeanSquaredError(MSE) = \frac{\sum_{t=1}^{n} e_t^2}{n} = \frac{\sum_{t=1}^{n} (y_t - \hat{y}_t)^2}{n} \qquad (4.2)$$

4.5 Box-Jenkins Methodology for Time Series Modeling

Decomposition of a time series data into its components, while being instructive and revealing, is a difficult job. Moreover, it causes greater errors by accumulation of component errors. To avoid these difficulties, Box and Jenkins (1976) developed a new methodology, which in essence, does the same job but unifies all concepts discussed above. In this method, using some transformations such as simple and seasonal differences the trends, seasonal and cyclical components present in the data are removed. Then, a family of models is entertained for the transformed data, which is expected to be as simple as possible. Having estimated the parameters of the model, the accuracy is checked and the model is appropriately modified. Iterating this process gives a model which fits the data well. This final model is used for prediction.

Essentially the Box-Jenkins procedure consists of four basic steps: tentative identification, estimation, diagnostic checking, and forecasting. Identification is based on the comparison of Sample Autocorrelation Function (SACF) and Sample Partial Autocorrelation Function (SPACF) of the time series data with those of known families of models. These families of models are autoregressives of order (AR) p = 1, 2,..., moving averages (MA) of order q = 1, 2,..., mixed autoregressive-moving averages of order (p, q), and autoregressive integrated moving averages of order (p, d, q), d = 0, 1, 2,..., where d is the degree of differencing to reach stationary. Estimation is done by exploiting the general method of estimation of the maximum likelihood.

Diagnostic checking is concerned with ascertaining the adequacy of the entertained model. If the model proves to be inadequate, it must be modified and improved. The diagnostic methods use the forecast errors to detect the flaws in the chosen model. They help in deciding the method of improving the model.

When a final model is determined, a model is used to compute the future values. These values are expressed in two forms: point estimates and interval estimates. The future values along with their respective probability limits are the products of time series modelling.

A point to be considered in forecasting future values of a positive variable is that negative lower limit for the forecast is obtained. In this situation, zero is used in place of a negative lower limit.

The Box-Jenkins approach is based on the notion of stationary time series briefly explained in the following section.

4.6 Stationary and Non-stationary Time Series

Classical Box-Jenkins models are used for stationary time series. Thus, to tentatively identify a Box-Jenkins model, it is necessary to verify that the time series used in forecasting is stationary. If it is not, the time series should be transformed into a series of stationary time series values.

Intuitively, a time series is called stationary if their statistical property, such as mean, variance remains essentially constant through time. If n values of $y_1, y_2, \ldots, y_n$ of a time series are observed, by examining their plot against time, their stationery can be checked. If n values seem to fluctuate with constant variation around a constant level, then it is reasonable to believe that the time series is stationary. In practice, this is done with the help of sample autocorrelation function (SACF) and sample partial autocorrelation function (SPACF). If n values do not fluctuate around a constant mean or do not fluctuate with constant variation, then it is reasonable to believe that the time series is nonstationary. In this case, one can sometimes transform the nonstationary time series values into stationary time series values, by taking the first, second or higher differences of the nonstationary time series values (Bendat and Piersol 1966).

The first difference of the time series values $y_1, y_2, \ldots, y_n$ are defined as:

$$Z_t = y_t - y_{t-1} = \nabla y_t \\ t = 2, \ldots, n \tag{4.3}$$

where, the difference operator ∇ is related to backward shift operator B, i

$$e.\nabla = 1 - B \tag{4.4}$$

where $By = y_{t-1}$ and consequently, $B_j y_t = y_{t-j}$.

4.7 The Sample Autocorrelation and Partial Autocorrelation Functions

The behavior of the SACF and the SPACF are important in tentative identification of stationary time series models. For the values of a stationary time series Z_b, Z_{b+1}, ..., Z_n which may be the original time series values or the transformed time series values, SACF is defined as follows. The sample autocorrelation at lag k denoted by r_k is:

$$r_k = \frac{\sum_{t=b}^{n-k}(z_t - \bar{z})(z_{t+k} - \bar{z}^c)}{\sum_{t=b}^{n}(z_t - \bar{z}^c)^2} \quad (4.5)$$

$$\bar{z} = \frac{\sum_{t=b}^{n} z_t}{n - b + 1}$$

Considering r_k a function of lag k, for k = 1, 2,..., K, it calls the sample autocorrelation function (SACF). This quantity measures the linear relationship between time series observations separated by a lag of k time units. The r_k is a coefficient of correlation and it is always between −1 and +1. The standard error of r_k is given by:

$$s_{r_k} = \frac{\left[\frac{1 + 2\sum_{j=1}^{k-1} r_j^2}{n-b+1}\right]}{2} \quad (4.6)$$

$$k = 1, 2, \ldots$$

The t_{r_k} statistic is then computed as:

$$t_{r_k} = \frac{r_k}{s_{r_k}} \quad (4.7)$$

which is used to test the significance of r_k, for k = 1, 2,...

Plotting r_k against k provides the SACF. The behavior of this function is a key tool for identification of the stationary of a time series and its order.

To employ the Box-Jenkins approach, one must examine and try to classify the behavior of the SACF. The SACF for a non-seasonal time series can display a variety of behaviors (Bowerman and O'Connell 1987). These are explained as follows:

- If the SACF of the time series values $Z_b, Z_{b+1}, \ldots Z_n$ either cuts off fairly fast or dies down fairly quickly, then this series should be considered as stationary.
- If the SACF of the time series dies down extremely slowly then the series should be considered as non-stationary.

Thus the procedure for finding a stationary time series is to compute SACF of the original series and then examine its behavior. If it is not stationary, by methods such as differencing or taking logs, the data is transformed. The SACF for the transformed series is computed and the new SACF is examined. The procedure is continued until stationarity is reached.

The SPACF is another important tool for identification of time series models. The sample partial autocorrelation at lag k is defined with:

$$k = 1 \Rightarrow r_{kk} = r_1$$

$$k = 2, 3, \ldots \Rightarrow r_{kk} = \left(r_k - \sum_{j=1}^{k-i} r_{(k-1),j} \cdot r_{k-j} \right) \Big/ \left(1 - \sum_{j=1}^{k-i} r_{(k-1),j} \cdot r_{k-j} \right) \quad (4.8)$$

where

$$r_{kj} = r_{(k-1)j} - r_{kk} r_{(k-1),(k-j)}$$
$$j = 1, 2, \ldots, (k-1) \quad (4.9)$$

The standard error of risk is defined as:

$$S_{r_{kk}} = \left[\frac{1}{(n-b+1)} \right]^{0.5} \quad (4.10)$$

And the student's t_{kk}—statistic is given by:

$$t_{r_{kk}} = t_{r_{kk}} / S_{r_{kk}} \quad (4.11)$$

The precise interpretation of the SPACF at lag k is rather complicated. However, this quantity can intuitively be thought of as the sample autocorrelation of time series observations, separated by a lag of k time units, with the effects of the intervening observations eliminated.

As with SACF, one must examine the behavior of SPACF and classify it to provide guidelines for identification of time series models: the SPACF can also display a variety of different behaviors. First the SPACF can cut off. This can occur when r_{kk} is not statistically significant beyond some number K. In general, K is at most equal to 3. In such a case $|t_{r_{kk}}|$ would be small and generally less than 2 for k < K. Second, the SPACF may die down if this function does not cut off but rather decreases in a "steady manner". This function may exhibit exponential decay, damped sine-wave or as a mixture of them, Bowerman and O'Connell (1987).

4.8 Classification of Non-seasonal Time Series Models

For a time series consisting of $Z_b, Z_{b+1}, \ldots, Z_n$, where Z_t is the original or transformed values of a time series, an autoregressive model of order $p, AR(p)$, is defined as:

$$Z_t = \phi_1 Z_{t-1} + \phi_2 Z_{t-2} + \cdots + \phi_r Z_{t-p} + a_t \qquad (4.12)$$

where $\phi_1 \ldots \phi_r$ are fixed coefficients and a_t, t = 1, 2,…, n are independent random variables with zero mean and constant variance σ_t^2. They are usually assumed as normally distributed. Using the backward shift operator B, Eq. (4.12) can be written as:

$$\phi_p(B)Z_t = a_t \qquad (4.13)$$

where $\phi_p(B) = 1 - \phi_1 B - \cdots - \phi_p B_p$ and $BZ_t = Z_{t-1}, \ldots, B_p Z_t = Z_{t-p}$.

A moving average model of order q, MA (q), is represented as:

$$Z_t = a_t - \theta_1 a_{t-1} - \cdots - \theta_q a_{t-q} \qquad (4.14)$$

Or employing the backward shift operator B,

$$Z_t = \theta_q(B) a_t \qquad (4.15)$$

With

$$\theta_q(B) = 1 - \theta_1 B - \cdots - \theta_q B_q \qquad (4.16)$$

The general nonseasonal autoregressive moving average model of order (p, q) is

$$Z_t = \delta + \phi_1 Z_{t-1} + \cdots + \phi_p Z_{t-p} + a_t - \theta_1 a_{t-1} - \cdots - \theta_q a_{t-q} \qquad (4.17)$$

This model utilizes a constant term δ. It has an autoregressive part which expresses the current value Z_t as a function of past values $Z_{t-1}, Z_{t-2}, \ldots, Z_{t-p}$ with unknown coefficients (parameters) $\phi_1, \ldots, \phi_p$. In addition, it has a moving average part which is represented by $a_t, a_{t-1}, \ldots, a_{t-q}$, with unknown fixed parameters $\mathbf{q_1}, \ldots, \mathbf{q_q}$. The variable Z_t is also considered as a function of a random variable, as, $a_{t-1}, \ldots, a_{t-q}$.

In Eq. (4.17), the constant term δ can be shown as equal to equal $\mu \phi_p(B)$, where μ is the mean of the stationary time series Z_t. In concise notation, Eq. (4.17) is presented as:

$$\phi_p(B) Z_t = \delta + \theta_q(B) a_t \qquad (4.18)$$

There are statistical tests, which can be used to decide whether to include δ in the model.

If the stationary time series $Z_b, Z_{b+1}, \ldots, Z_n$ is the original series, then assuming μ is equal to zero, this implies that these original time series values are fluctuating around a zero mean, whereas $\mu \neq 0$ implies that these original values are fluctuating around a non-zero mean. In such a case one can use $Z_t - \bar{Z}$ in place of Z_t. Then δ can be removed from the model. If the stationary time series $Z_b, Z_{b+1}, \ldots, Z_n$ are different from those of the original time series values, where μ is not assumed to be zero, it can be assumed that there is a deterministic trend in those original values. Here the deterministic trend refers to a tendency to the original values to move persistently upward (if $\delta > 0$) or downward (if $\delta < 0$). If a time series does not exhibit a deterministic trend, then any trend (or failure of the series to fluctuate around a central value) is stochastic. The stochastic trend is more realistic in practical situations since it does not dictate a certain path to be taken by the future values.

4.9 Guidelines for Choosing a Non-seasonal Models

ARMA (p, q) models of Eq. (4.17) are specified by choosing suitable orders for AR operator ϕ_p (B) and MA operator θ_q (B). This boils down to specifying p and q as positive integers. It is illustrated by some guidelines for choosing such numbers. See Bowerman and O'Connell (1987).

4.10 Seasonal Box-Jenkins Models

Seasonality may be defined as the common feature of most time series data being the periodic pattern of fluctuations in time series values. For example the meteorological time series recorded in a location such as temperature, rainfall, and radiation, exhibit a marked periodic behavior of 12 months. River discharges have periodic nature too. This feature can be accompanied by any one or more of a trend, cyclical and irregular fluctuations. Seasonal change is an example of non-stationarity, which can be removed by a seasonal differentiation. Seasonal differencing takes the differences of two similar observations, one from each period. For example in the case of monthly average temperature, the difference of the values of the same months from consecutive years removes seasonality. For seasonal differencing, the seasonal operator ∇s is defined as:

$$\nabla_s = 1 - B^s \tag{4.19}$$

4.10 Seasonal Box-Jenkins Models

where S is the length of season, for instance S = 4 for quarterly data and S = 12 for monthly data. Let y_t^* denote an appropriate pre-differencing transformation. This can be shown as, $y_t^* = \ln y_t$ if a logarithmic transformation is needed. Another example is the normalizing transformation of Box and Cox (1964) which is defined as:

$$y_t = \begin{cases} \frac{y^\lambda - 1}{y^{*\lambda}} & \lambda \neq 0 \\ \ln y, & \lambda = 0 \end{cases} \quad (4.20)$$

where $y^{*\lambda}$ is the geometric mean of y^λ values.

Then the general stationery transformation is given by:

$$Z_t = \nabla_S^D \nabla^d y_t^* = (1 - B^s)^D (1 - B)^d y_t^* \quad (4.21)$$

where d is the degree of non-seasonal differencing and D is the degree of seasonal differencing used to reach stationary. Either of D and d can be taken as 0, 1, 2 or at most 3.

The SACF within each season behaves as it was described for non-seasonal models. Ignoring the behavior of the SACF within each season and only considering it at lag's multiples of S can describe the seasonal behavior of the time series.

Similarly, the SPACF of seasonal models can be studied within and between seasons. Once the stationarity transformation is performed, there is:

$$Z_t = (1 - B^s)^D (1 - B)^d y_t^* \quad (4.22)$$

which provides $Z_b, Z_{b+1}, \ldots, Z_n$ model describing these values. This model consists of two components determined by their respective operators. One set of operator's models the seasonal pattern while the other set does the non-seasonal pattern, of the data. The general model of order (p, P, q, Q) is written as:

$$\phi_p(B)\Phi_P(B^s)Z_t = \delta + \theta_q(B)\Theta_Q(B^s)a_t \quad (4.23)$$

where $\phi_p(B) = 1 - \phi_1 B - \cdots - \phi_p B_p$ is the non-seasonal autoregressive operator of order p, $\Phi_P(B^s) = 1 - \Phi_{1,s}B^s - \Phi_{2,s}B^{2s} - \cdots - \Phi_{P,s}B^{Ps}$ is called the seasonal autoregressive operator of order P, $\theta_q(B) = 1 - \theta_1(B) - \cdots - \theta_q B^q$ is the non-seasonal moving average operator of order q, $\Theta_Q(B^s) = 1 - \Theta_{1,s}B^s - \Theta_{2,s}B^{2s} - \cdots - \Theta_{Q,s}B^{Qs}$ is called the seasonal moving average operator of order Q, and δ is a constant term. $\phi_1, \phi_2, \ldots, \phi_p; \Phi_{1,s}, \Phi_{2,s}, \ldots, \Phi_{P,s}; \theta_1, \theta_2, \ldots, \theta_q; \theta_{1,s}, \theta_{2,s}, \ldots, \theta_{Q,s}$ and δ are unknown parameters which can be estimated from sample data. $a_t, a_{t-1}, \ldots$ are random shocks which are assumed to be statistically independent of each other, and identically distributed as normal with zero mean, and a constant variance. This is true for each and every time period t.

4.11 Guidelines for Identification of Seasonal Models

As in case of non-seasonal identification of the seasonal model (Eq. (4.23)), requires whether to include δ in the model. This task is accomplished by performing a significant test using $\bar{Z}$.

The behavior of the SACF and PSACF of the $Z_b, Z_{b+1}, \ldots, Z_n$ values at the non-seasonal level is used to decide which (if any) of the non-seasonal moving average operators of order q, and $\theta_q(B) = 1 - \theta_1(B) - \cdots - \theta_q B_q$ and the non-seasonal autoregressive operator of order p, and $\phi_p(B) = 1 - \phi_1 B - \cdots - \phi_p B_p$ should be employed.

The behavior of the SACF and the SPACF of $Z_b, Z_{b+1}, \ldots, Z_n$ values at the seasonal level are used to determine which (if any) of the seasonal moving average operators of order Q, and $\Theta_Q(B^s) = 1 - \Theta_{1,s} B^s - \Theta_{2,s} B^{2s} - \cdots - \Theta_{Q,s} B^{Qs}$ and the seasonal autoregressive operator of order P, and $F_P(B^s) = 1 - F_{1,s} B^s - F_{2,s} B^{2s} - \cdots - F_{P,s} B^{Ps}$ should be utilised. In this regard, some guidelines are available which are used to specify a tentative model, see Bowerman and O'Connell (1987).

Having identified the appropriate operators, and inserting the appropriate stationarity transformation, one obtains the model.

$$\phi_p(B)\phi_p(B^s)\nabla_S^D \nabla^d y_t^* = \delta + \theta_q(B)\theta_Q(B^s)a_t \qquad (4.24)$$

whose parameters should be estimated and its accuracy should be checked. These steps are similar to those of non-seasonal models.

4.12 Diagnostic Checking

Diagnostic checking is used to see whether or not the identified and estimated model is adequate. If the model proved to be inadequate, it must be modified and improved. The diagnostic methods employed will help to decide how the model can be improved.

A good way of finding the adequacy of an overall model is to analyze the residuals obtained from the model. According to the assumptions, which the model is based upon, the residuals should be independent and identically distributed. They should follow a normal distribution. Thus, the diagnostic examination methods consist of checking for independence, normality, and the absence of any systematic behavior in the residuals.

Before embarking on a specific test, an overall check, which utilizes the sample autocorrelation function of the residuals, is employed. This is called a portmanteau lack of fit test. The test statistic for the portmanteau test is either the Box-Pierce (Box and Jenkins 1976) Statistic $Q = n' \sum_{i=1}^{K} r_i^2(\hat{a})$ or the preferred Ljung-Box (Box and Jenkins 1976) statistic $Q = n'(n'+2) \sum_{i=1}^{K} r_i^2(\hat{a})/(n'-i)$. where $n' = n - (d - SD)$

4.12 Diagnostic Checking

in which n is the number of observations in the original time series, and S is the length of the season in the series. A good choice of K is $2\sqrt{n'}$. These statistics both have χ^2 distribution with v degrees of freedom, where $v = K - m_p$ and m_p are the number of parameters that must be estimated in the model under consideration. Large values of Q or Q^* signify a lack of fitting the model, whose sources can be traced by the following specific tests.

Independence of residuals could be checked by various routine statistical methods. However, this is best carried out by sample autocorrelation function of the residuals (RSAC). If independence holds, the RSAC would reflect it by having no significant correlations.

Normality could also be checked by several methods but the easiest one is the *p-p* plot (normal probability plot) of the residuals. In this method, a plot is drawn in which the x-axis indicates the percentiles of the standard normal distribution and y-axis represents the observed percentiles of the standardized residuals. If residuals follow a normal distribution, the *p-p* plot would resemble a straight line from (0, 0) to (1, 1). Wild departures from this line are a sign of non-normality and may need some sort of transformation for the data to make them normal.

To ascertain the absence of any systematic pattern in the residuals, especially of the periodic nature, the integrated paradigm can be used. The integrated (cumulative) periodogram for a random series having no periodic feature is a straight line of 45° passing through (0, 0) and (0.5, 1). A significant departure from this line suggests some periodic aspects left in the residuals, which could be retrieved and incorporated, in the model.

In general a flow diagram of the iterative approach for the Box-Jenkins model building and stages to reach an adequate model is illustrated in Fig. 4.2, and Table 4.1.

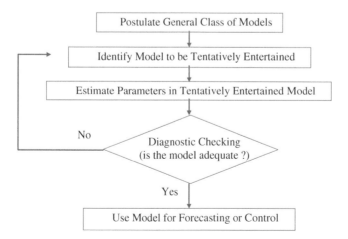

Fig. 4.2 Stage in the iterative approach to model building (Box and Jenkins 1976)

Table 4.1 Stage of Box-Jenkins modeling (Lohani and Wang 1987)

Step	Description
1	Check the data for normality
	No transformation
	Square root transformation
	Logarithmic transformation
	Power transformation
2	Identification
	Plot of the transformed series
	Autocorrelation function (ACF)
	Partial autocorrelation function (PACF)
3	Estimation
	Maximum likelihood estimate (MLE) for the model parameters (Ansley algorithm)
4	Diagnostic checks
	Over-fitting
	Examination of residuals (modified Portmanteau test)
5	Model Structure Selection Criteria
	(a) AIC criteria
	(b) PP criteria
	(c) BIC criteria

4.13 Exponential Smoothing Methods

Exponential smoothing is a forecasting technique that attempts to track changes in a time series. This is done by using newly observed values to update the estimates of parameters in the time series model. The well-known methods of exponential smoothing are:

1. Simple exponential smoothing,
2. Winter's Method appropriates for seasonal data, and
3. One- and two- parameter double exponential smoothing.

When using an exponential smoothing method, it is often useful to employ adaptive control procedures to monitor the accuracy of the forecasting system (Chow 1965).

Exponential smoothing techniques are essentially equivalent to some special Box-Jenkins models (McKenzie 1984). For this reason, only a brief mention of their main features is presented (Bowerman and O'Connell 1987).

4.13.1 Simple Exponential Smoothing

For a stationary series, suppose:

$$y_t = \beta_0 + \varepsilon_t \tag{4.25}$$

In general, if $a_o(T)$ is the estimate in time period T for the average level β_0 of Eq. 4.25 then, a point forecast τ steps ahead made in time period T for $y_{T+\tau}$, is given by:

$$\hat{y}_{T+\tau}(T) = a_o(T) \tag{4.26}$$

To start the forecasting process, $a_o(0)$ is needed which is usually the average of the first one third or half of historical data, depending on the length of the series. That is:

$$a_o(0) = \sum_{t=1}^{[n/k]} y_t / [n/k] \tag{4.27}$$

where $k = 2$ or 3 and $[n/k]$ is the integral part of n/k. The portion used to estimate $a_o(0)$ should not be too long or too short.

The updating equation for $a_o(T)$ is given by:

$$a_o(T+1) = \gamma y_{T+1} + (1-\gamma)a_o(T) \tag{4.28}$$

where γ is the smoothing constant, often taking a value between 0.01 and 0.30 in most applications. Its best value can be obtained by trial and error via minimizing the sum of squared forecast errors. However, the guiding principle is that the smaller value of γ indicates that the average level of the time series does not change much over time.

Equations 4.26 and 4.28 allow the computation the forecasting of the values for the desired time points.

The expression for constructing interval forecasts, of giving confidence level, for $y_{t+\tau}$ in time period T, is given by:

$$[\hat{y}_{t+\tau}(T) - B_{t+\tau}^{[100(1-x)]}(T), \hat{y}_{T+\tau}(T) - B_{T+\tau}^{[100(1-x)]}(T)] \tag{4.29}$$

where, $B_{t+\tau}^{[100(1-x)]}(T) = Z_{\alpha/2} 1.25 \Delta(T)$ and $\Delta(T)$ is the average absolute forecast error for the time period T, i.e.

$$\Delta(T) = \sum_{t=1}^{T} |y_t - a_o(t-1)|/T \tag{4.30}$$

The value of $Z_{\alpha/2}$ is the $100(1 - \alpha/2)$ the centile of standard normal distribution.

Moreover, considering the time series value y_{T+1}, one can update $a_o(T)$ to $a_o(T+1)$ by Eq. 4.28, and also update $\Delta(T)$ to $\Delta(T+1)$ by:

$$\Delta(T+1) = \frac{T\Delta(T) + |y_{T+1} - a_o(T)|}{T+1} \tag{4.31}$$

Now, using Eqs. 4.28 and 4.31 in Eqs. 4.26 and 4.29, the update forecast is obtained.

4.14 Winter's Method

Winter's method is an exponential smoothing procedure appropriate for seasonal data (Winters 1960). Winter's multiplicative model is given by:

$$y_t = (\beta_0 + \beta_1 t)SN_t + \varepsilon_t \tag{4.32}$$

That, this model assumes a linear trend and a multiplicative seasonal variation. In order to apply this method, the following calculations are required:

1. An initial estimate $b_1(0)$ of β_1
2. An initial estimate $a_o(0)$ of β_0
3. An initial estimate $s_n t(0)$ of SN_t

From the historical data of the last m year, an average $\bar{y}_i$ is defined (for the i th year, $i = 1, 2, \ldots, m$. then $b_1(0) = (\bar{y}_m - \bar{y}_1)/(m-1)s$. s is the length of season), the initial estimate for β_0, the average level of the series at $t = 0$ is given by:

$$a_o(0) = \bar{y}_1 - \frac{s}{2} b_1(0) \tag{4.33}$$

The initial estimate for the s seasonal factors is given by:

$$s_t = y_t / \{\bar{y}_i - [(s+1)/2 - j]b_1(0)\} \tag{4.34}$$

where $\bar{y}_i$ is the average of the observations for the year in which season t occurs (if $1 \leq t \leq s$, then $i = 1$, if $s + 1 \leq t \leq 2s$, then $i = 2$, etc.).

The letter j represents the position of season t within the year. For monthly data, $j = 1$ represents January, $j = 2$ is February and so forth. [S_t must be computed for each season (month, quarter, etc.) t occurring in the year 1 through m].

Equation 4.32 yields m values for S_t. These m distinct estimates for seasonal factor are averaged to give

$$s\bar{n}_t = \frac{1}{m} \sum_{k=0}^{m-1} S_{t+ks} \tag{4.35}$$

$$t = 1, 2, \ldots, s$$

4.14 Winter's Method

which is the average seasonal index for each different season. Finally, the initial estimate $sn_t(0)$ of SN_t is given by:

$$sn_t(0) = s\bar{n}_t \left[\frac{S}{\sum_{t=1}^{S} s\bar{n}_t} \right] \qquad (4.36)$$

$$t = 1, 2, \ldots, s$$

The updating equations are defined as:

$$a_o(T) = \gamma \frac{y_T}{sn_T(T-s)} + (1-\gamma)[a_o(T-1) + (T-1)] \qquad (4.37)$$

where γ is a smoothing constant, $0 < \gamma < 1$.

$$b_1(T) = \theta[a_0(T) - a_0(T-1)] + (1-\theta)b_1(T-1) \qquad (4.38)$$

where θ is a smoothing constant, $0 < \theta < 1$.

$$sn_t(T) = \omega \frac{y_T}{a_0(T)} + (1-\omega)sn_T(T-S) \qquad (4.39)$$

where ω is a smoothing constant, $0 < \omega < 1$.

Having the updated values for the components of Eq. (4.32), which are given in Eqs. 4.37–4.39, a point forecast made at time T for $y_{T+\tau}$ is obtained by:

$$\hat{y}_{T+\tau}(T) = [a_0(T) + b_1(T)\tau]sn_{T+\tau}(T+\tau-s) \qquad (4.40)$$

Interval forecasts based on complicated formulas can be constructed which can be found in Bowerman and O'connell (1987).

4.15 One and Two-Parameter Double Exponential Smoothing

The two parameter exponential smoothing is a special case of Winter's method where SN_t equated to unity for all values of t. That is, $y_t = b_0 + b_1 t + e t$. Thus, the previous results in the Winter's method also apply in this case.

In one-parameter of exponential smoothing, the smoothing constants are related to each other, hence one parameter suffices. Thus let $\gamma = 1 - w^2$ and $\theta = \frac{2w}{1+w}$ where $w = 1 - d$, and δ is a smoothing constant which is chosen to lie between the values of 0 and 1. Therefore;

$$\hat{y}_{T+\tau}(T) = a_0(T) + b_1(T)\tau \qquad (4.41)$$

4.16 Adaptive Control Procedures

One cannot expect a forecasting system to produce perfect forecasts of a time series. Therefore, it is required to determine whether the forecast errors are tolerable or are beyond the reasonable bounds. To this end, assume a history of T one-step ahead forecast errors, $e_1(\gamma), e_2(\gamma), \ldots, e_T(\gamma)$ is computed. Here, γ denotes the fact that the forecast errors are dependent on the smoothing constant γ, used to compute a one-step ahead forecast. The cumulative forecast error can be defined by $C(\gamma, T) = \sum_{t=1}^{T} e_t(\gamma)$ until time period T is reached. Then $C(\gamma, T)$ will be equal to $C(\gamma, T-1) + eT(\gamma)$.

Now, the mean absolute deviation is defined by:

$$D(\gamma, T) = \frac{\sum_{t=1}^{T} |e_t(\gamma)|}{T} \tag{4.42}$$

Then the tracking signal $TS(\gamma, t)$ is defined as:

$$TS(\gamma, T) = \left| \frac{C(\gamma, T)}{D(\gamma, T)} \right| \tag{4.43}$$

If $TS(\gamma, T)$ is "large", it means that $C(\gamma, T)$ is also large relative to $D(\gamma, T)$. Equivalently this reflects that absolute errors are large and some measures should be taken. These measures can be taken as:

1. Change the smoothing constant,
2. Change the model,
3. Use different values of smoothing constant at different times.

Some procedures have been developed to implement these measures on the computer (Chow 1965).

Considering the deficit of water in Iran, protection of water resources against pollution is vital. In this regard, water quality monitoring is a tool which produces up to date information. Having a great amount of raw data without interpretation is not sufficient, and it is necessary to analysis data and predicts the variation of water quality in the future for any decision making on water quality management. Recently, more researchers have become interested in the application of time series models for the prediction of water quality. Time series approach for analyzing water resources were first applied by Thomann (1967) who studied variation of temperature by the time and dissolved oxygen level for the Delaware Estuary. The data was obtained by continuous recording by monitoring stations, operated jointly by the U.S. Geological Survey Department and the city of Philadelphia. Carlson et al. (1970) and McMichael and Hunter (1972) reported the successful use of the Box-Jenkins method for time series analysis.

The Box-Jenkins method for the time series analysis was applied to model the hourly water quality data recorded in the St. Clair River near Corunna, Ontario, for chloride and dissolved oxygen levels by Huck and Farquhar (1974), the models which were obtained physically reasonable and successful. Autoregressive and first difference moving average models represented the chloride data well. Lohan and Wang (1987) also reported to have used this model to study the monthly water quality data in the Chung Kang River located at the northern part of Miao-Li County in the middle of Taiwan. Jayawardena and Lai (1991) applied an adaptive ARMA model approach for water quality forecasting. MacLeod and Whitfield (1996) analyzed water quality data using Box-Jenkins time series analysis of the Columbia River at Revelstoke. Caissie et al. (1998) studied water temperature in the Catamaran Brook stream. The short-term residual temperatures were modeled using different air to water relations, namely a multiple regression analysis, a second-order Mar for process, and a Box-Jenkins time series model. Asadollahfardi (2002) applied Box Jenkins and. Exponential smoothing models to monthly surface water quality data in Tehran for 3 years. Most of the models indicated seasonality. Kurunc et al. (2004) applied ARIMA and Thomas- Fiering techniques for 13 years to monthly data of the Duruacasu station at Yesilirmark River. Hasmida (2009) applied ARIMA model (parametric method) and Mann-Kendall test (non-parametric method) to analyze the water quality (NH4, turbidity, color, SS pH, Al, Mn and Fe.) and rainfall-runoff data for Johor River recorded for a long period (2004 to 2007). He described that all of the water quality parameters were generated by ARIMA processes ranges from ARIMA (1,1,1) to (2,1,2). He concluded that color, Turbidity, SS, NH4 and Mn follow a similar trend with the rainfall-runoff pattern while pH, Al and Fe have the opposite trend compare to rainfall-runoff pattern.

4.17 Application of Time Series

4.17.1 A Case Study: Latian Dam Water Quality

Time series models were applied to some parameters of inlet and outlet water quality in Latian dam, Tehran. There are five water quality monitoring stations downstream and upstream of the dam. Among which there are remarkably significant because of passing of the greatest volume of water (Fig. 4.1). These stations are Roudak on Jadjrud River, Aliabad on the Lavark River and Zir-e-pol on the outlet of the dam. Table 4.2 and Fig. 4.3 show the situation and characteristics of the dam and the stations.

The study area is a 71,000 hectare river basin in the Alborz Mountains. The rainfall regime is primarily derived from the Mediterranean region. According to pluviometry data of 14 stations in the region, annual rainfall variations in height with a 20-year statistical period follows the equation below:

$$P = -185.3 + 0.379Z \tag{4.44}$$

where, Z is height above sea level and P is the annual rainfall (Kakavand 2001).

Table 4.2 Situation and characteristics of stations upstream and downstream of Latian dam

River	Station	Longitude (degree/min)	Latitude (degree/min)	Altitude (m)	Basin area (km²)
Jadj Rud	Roodak	51°33'	35°51'	1,690	416
Lavark	Ali Abad	51°41'	35°48'	1,600	103
Afjah	Narvan	51°40'	35°50'	1,750	30
Galandovak	Najar Kola	51°38'	35°49'	1,700	59
JadjRud	Zir-e-pol	51°41'	35°47'	1,560	710

In the catchment area, annual average temperature is 10 °C. The hottest month of the year is late of July to late of August with a maximum temperature of 34 °C and the coldest is late of December to late of January with the minimum temperature of −8 °C. Average rainfall of Latian basin is more than 500 mm a year. Latian Dam is located at 35°47' N, 51°40' E. In addition to producing 70,000 MW/hours hydropower energy, it supplies drinking water to some parts of Tehran and also agricultural water to some parts of the Southeastern part of Tehran (Varamin Plain). Some characteristics of the dam are shown in Table 4.2.

This case study was primarily aimed at developing suitable and confident time series models for water quality data in two inlets and an outlet of the dam. The second objective was to predict variations in water quality from developed models to be used in water quality management. Third aim is to show application of time series in water quality (Asadollahfardi et al. 2012).

4.17.1.1 The Software

Statistical Analysis System (SAS) version9/1 was applied for calculations and analyze of the models of this case study. This software needs to be programmed; however, there are also some menus for simplicity. First, it is necessary to build a library in the software to save data and calculations of each stage. Figure 4.4 shows the procedures for building, confirming and, forecasting models with SAS software.

The aim of this case study was to develop proper models for $Ca^{++}, Mg^{++}, SO_4^{--}, pH, HCO_3^-, Na^+, Cl^-$ and TDS parameters. The data used in building time series model as well as confirming and comparing the models were monthly collected for 24 years (1981–2005) by local water authorities in Tehran. For validation of the models, the predicted monthly values from September till March 2005 were compared to the observed values. Finally, a relation was made for every parameter (Asadollahfardi et al. 2012). The results are presented in Tables (4.4), (4.5) and (4.6).

As shown in Tables 4.4 and 4.5 most of the models developed for water quality in Aliabad and Roudak stations were Auto Regressive Integrated Moving Average

4.17 Application of Time Series

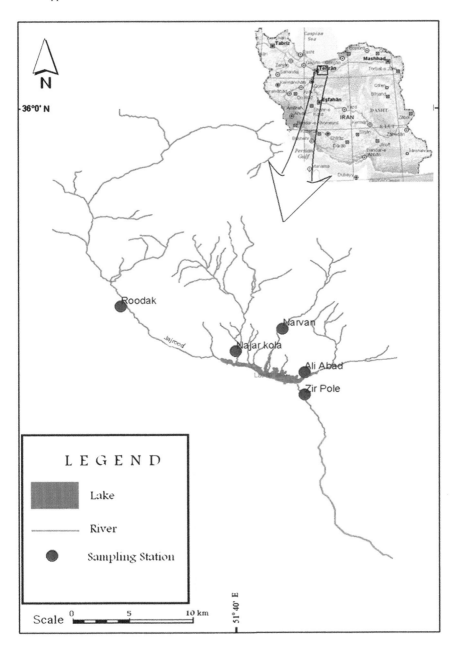

Fig. 4.3 Location of water quality on Latian Dam

(ARIMA) type; however, for Zir-e-pol station, there were seasonal models as well as non- seasonal ones. The obtained values were as follows: ARIMA models with autoregressive order 2 and seasonal autoregressive order one for Na^+, Mg^{++} and

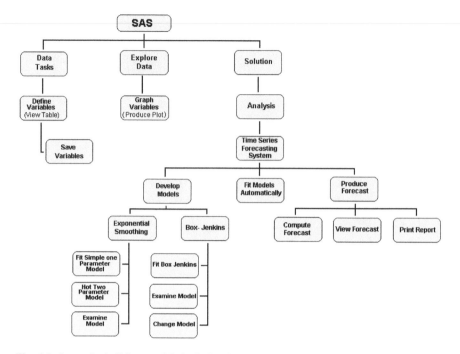

Fig. 4.4 Stages for building models in SAS software

Table 4.3 Characteristics of Latian Dam

Type of dam	Concrete and weight
Height from foundation	107 m`
Height from riverbed	80 m
Length of crest	450 m
Total capacity of reservoir	95×10^6 m^3
Useful capacity of reservoir	85×10^6 m^3
Capacity of evacuation of spillways	Uncovered 650 m^3
	Tunnel 1,100 m^3

SO_4^{--} parameters; as non-seasonal ARIMA with autoregressive order two for Cl^- pH, Ca^{++} and seasonal ARIMA with autoregressive order 1 and moving average order one for TDS and HCO_3^- non. Some of the models indicated in Tables 4.3, 5.4 and 6.4, will be discussed in detail in the following section.

If p-Value is more than 0. 9, the model has proven to be excellent, from 0.75 to 0.9 it is evaluated as good and p-value between 0.5 and 0.75 indicates that the model is average.

4.17 Application of Time Series

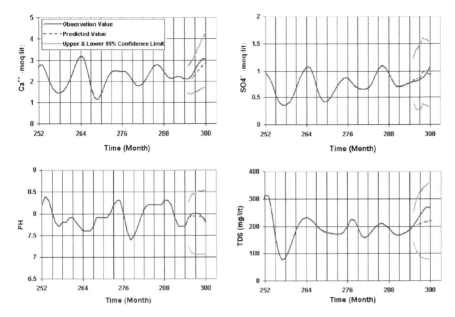

Fig. 4.5 Diagram of time series for each of water quality parameters and their predictions (the predictions are based on 1 month ahead projections)

4.17.1.2 Selected Models of Few Water Quality Parameters in Zir-e-Pol Station

Ca^{++}

As described in Fig. 4.5 the proper model for calcium (Ca^{++}) was ARIMA. (2,0,0)(0,0,0) The relation of the model is as follows:

$$Z_t = 2.186 + 1.02Z_{t-1} - 0.338Z_{t-2} + a_t \qquad (4.45)$$

where Z_t is the amount of calcium, and a_t stands for error. The standard error is 0.331 according to comparison approaches. Akaike Information Criteria (AIC) and Schwartz Bayesian Information Criteria are less than the other models. Also correlation coefficient is 0.95 which is proper.

p-Value is 0.99 which indicates that the model is excellent (Table 4.6).

SO_4^{--}

Figure 4.5 shows variations of sulfate parameter and the best model for SO_4 is an ARIMA (2,0,0) (1,0,0) S with seasonal components (Table 4.4).

Table 4.4 Characteristics of developed models for Roodak station at Jajroud river (Asadollahfardi et al. 2012)

	Na^+	Ca^{++}	Mg^{++}	Cl^-
Suggested Model	ARIMA(1,0,0)(0,0,0)	ARIMA(1,0,0)(0,0,0)	ARIMA(2,0,0)(0,0,0)	ARIMA(2,0,0)(0,0,0)
Equation	$Z_t = 0.44 + 0.686(Z_{t-1}) + a_t$	$Z_t = 2.24 + 0.586(Z_{t-1}) + a_t$	$Z_t = 0.912 + 0.635(Z_{t-1}) - 0.188(Z_{t-2}) + a_t$	$Z_t = 0.295 + 0.662(Z_{t-1}) - 0.163(Z_{t-2}) + a_t$
AIC	−355.33	320.57	73.23	−425.29
SBC	−347.92	327.98	84.34	−410.47
R-Square	0.79	0.97	0.59	0.69
Std. Error	0.1332	0.411	0.2718	0.1182
Intercept	0.4367	2.2394	0.912	0.295
T	17.964	12.481	32.223	14.335
Prob > ITI	<.0001	<.0001	<.0001	<.0001
AR Lag1	0.6858	0.5859	0.6346	0.6619
T	16.328	39.183	11.1097	11.539
Prob > ITI	<.0001	<.0001	<.0001	<.0001
AR Lag2	–	–	−0.1883	−0.1632
T	–	–	−3.296	−2.854
Prob > ITI	–	–	0.001	0.004
SAR Lag12	–	–	.	.
	Na^+	Ca^{++}	Mg^{++}	Cl^-
T	–	–	.	.
Prob > ITI	–	–	.	.
P-Value	0.99	0.99	0.99	0.99
RV	0.0177	0.1692	0.0738	0.0139
ME	E	E	E	E
	HCO_3^-	SO_4^{--}	pH	TDS
	ARIMA(1,0,0)(1,0,0)	ARIMA(1,0,0)(0,0,0)	ARIMA(1,0,0)(0,0,0)	ARIMA(2,0,0)(0,0,0)

(continued)

4.17 Application of Time Series

Table 4.4 (continued)

	Na^+	Ca^{++}	Mg^{++}	Cl^-
Suggested Model				
Equation	$Z_t = 2.561 + (0.608(Z_{t-1}) + a_t) \times (0.145 (Z_{t-12}) + \varepsilon_t)$	$Z_t = 0.759 + 0.453(Z_{t-1}) + a_t$	$Z_t = 7.793 + 0.676(Z_{t-1}) + a_t$	$Z_t = 213.05 + 0.675(Z_{t-1}) - 0.113(Z_{t-2}) + a_t$
AIC	296.29	45.336	136.52	3034.66
SBC	307.4	52.743	143.92	3045.77
R-Square	0.89	0.79	0.86	0.95
Std. Error	0.394	0.259	0.302	37.826
Intercept	2.561	0.759	7.793	213.05
T	38.1248	27.718	145.29	42.818
Prob > |T|	<0.0001	<.0001	<.0001	<.0001
AR Lag1	0.6084	0.453	0.676	0.675
T	13.311	8.793	15.924	11.711
Prob > |T|	<0.0001	<.0001	<.0001	<.0001
AR Lag2	.	–	.	−0.1125
T	–	–	.	−1.953
Prob > |T|	–	–	.	0.05
SAR Lag12	0.145	–	.	–
T	2.5098	–	.	–
Prob > |T|	0.012	–	.	–
P-Value	0.99	0.99	0.99	0.95
RV	0.1552	0.0675	0.0915	1430.88
ME	E	E	E	E

LSW Level smoothing weight, *SL* Smoothed level, *AR* Autoregressive, *SAR* Seasonal autoregressive, *MA* Moving average, *RV* Residual variance (sigma squared), *ME* Model evaluation, *E* Excellent, *SBC* Schwartz Bayesian information criteria, *AIC* Akaike information criteria, Cl^- Chloride ion, HCO_3^- Bicarbonate ion, SO_4^{--} Sulfate ion, Ca^{++} Calcium ion, Mg^{++} Magnesium ion, Na^+ Sodium ion, *TDS* Total dissolved solids, *Std. Error* Standard Deviation

The equation of the model is as follow:

$$Z_t = 0.88 + (0.997Z_t - 1 - 0.381Z_{t-2} + a_t)(0.207Z_t - 12 + \varepsilon_t) \quad (4.46)$$

In this equation, part of $(0.997Z_t - 1 - 0.381Z_{t-2} + a_t)$ is non-seasonal autoregressive component of the model and $(0.2076Z_t - 12 + \varepsilon_t)$ is its seasonal autoregressive component. The standard error is 0.202. Akaike Information Criteria (AIC) and Schwartz Bayesian Information Criteria (SBC) of the model are less than other suggested models. The correlation coefficient is 0.53 (Table 4.6). The risk is less than 0.0001 and confidence level p-Value is 0.99 which proves the model to be excellent (Asadollahfardi et al. 2012).

Acidity (pH)

The best developed model for pH parameter was ARMA (2,0,0) (0,0,0) (Fig. 4.5). The equation for acidity is as follows:

$$Z_t = 7.775 + 0.842Z_{t-1} - 0.142Z_{t-2} + a_t \quad (4.47)$$

According to comparison methods, standard error of the model is 0.267. The AIC and the SBC of the model are less than other developed models. The correlation coefficient is 0.83 which is proper (Table 4.4). According to given assumptions, the amount of risk is less than 0.0297 and a confidence level of p-Value equals 0.97.

Total Dissolved Solid (TDS)

In Fig. 4 variations of TDS are shown. The best model developed for TDS is an ARIMA (1,0,1) (0,0,0) with autoregressive order one component and moving average order one. The equation of the model is as follows:

$$Z_t = 228.8 + 0.8Z_{t-1} + a_t + 0.591a_{t-1} \quad (4.48)$$

The standard error of the model is 29.36 and AIC and SBC are less than other developed models; the correlation coefficient is 0.91 which is proper (Table 4.4). The amount of the risk is 0.0001 and p-Value is 0.99 which evaluates the model as excellent.

The characteristics of all the models are shown in Tables 4.4, 4.5 and 4.6. It should be noted that there is no negative value in practice and 95 % confidence level considered in the calculation has caused the lower limit to be negative. Hence, negative values should be omitted or replaced by zero.

4.17 Application of Time Series

Table 4.5 Characteristics of developed models for Ali-Abad station in Lavark river (Asadollahfardi et al. 2012)

	Na^+	Ca^{++}	Mg^{++}	Cl^-
Suggested model	ARIMA(2,0,0)(0,0,0)	ARIMA(1,0,0)(0,0,0)	Simple Exponential Smoothing	ARIMA(2,0,0)(0,0,0)
Equation	$Z_t = 1.13 + 0.829 (Z_{t-1}) - 0.189 (Z_{t-2}) + a_t$	$Z_t = 2.66 + 0.647 (Z_{t-1}) + a_t$	$S_t = 0.999 Y_{t-1} + (1 - 0.999) S_{t-1}$ in above equation: S Prediction and Y Observation	$Z_t = 0.869 + 0.917(Z_{t-1}) - 0.153(Z_{t-2}) + a_t$
AIC	634.93	712.51	−232.591	328.59
SBC	649.66	723.56	−228.88	343.33
R-square	0.86	0.88	0.72	0.95
Std. error	0.7066	0.8079	0.0409	0.4195
LSW	.	.	0.999	.
T	.	.	24.4298	.
Prob > ITI	.	.	<.0001	.
SL	.	.	0.69	.
Intercept	1.13018	2.6573	–	0.86873
T	7.7518	15.799	–	7.555
Prob > ITI	<.0001	<.0001	–	<.0001
AR Lag1	0.82969	0.64663	–	0.91671
T	14.1942	14.4323	–	15.807
Prob > ITI	<.0001	<.0001	–	<.0001
AR Lag2	−0.18931	.	–	−0.15305
T	−3.2364	.	–	−2.6354
Prob > ITI	0.00135	.	–	0.0088
MA1	.	.	–	.
T	.	.	–	.
Prob > ITI	.	.	–	.

(continued)

Table 4.5 (continued)

	Na$^+$	Ca^{++}	Mg^{++}	Cl$^-$
SAR	.	.	–	.
Lag12				
T	.	.	–	.
Prob > |T|	.	.	–	.
P-value	0.99	0.99	0.99	0.99
RV	0.4994	0.6528	0.45903	0.1759
ME	E	E	E	E
	HCO$_3^-$	SO$_4^{--}$	pH	TDS
Suggested model	ARIMA(2,0,0)(0,0,0)	ARIMA(1,0,0)(0,0,0)	ARIMA(2,0,0)(0,0,0)	ARIMA(2,1,1)(1,0,0)
Equation	$Z_t = 2.45 + 0.987(Z_{t-1}) - 0.228(Z_{t-2}) + a_t$	$Z_t = 1.703 + 0.7(Z_{t-1}) + a_t$	$Z_t = 7.775 + 0.842(Z_{t-1}) - 0.142(Z_{t-2}) + a_t$	$Z_t = -0.615 + (0.955(Z_{t-1}) - 0.236(Z_{t-2}) + a_t - 0.995(a_{t-1})) \times (0.1499 Z_{t-12} + \varepsilon_t)$
AIC	455.02	914.61	101.15	3549.35
SBC	469.76	925.66	115.88	3571.43
R-square	0.97	0.83	0.77	0.91
Standard error	0.52	1.139	0.285	101.74
LSW	.	.	.	–
T	.	.	.	–
Prob > |T|	.	.	.	–
SL	.	.	.	–
Intercept	2.447	1.703	7.775	−0.615
T	17.4681	6.4469	118.086	−2.0069
Prob > |T|	<.0001	<.0001	<.0001	0.0457

(continued)

4.17 Application of Time Series

Table 4.5 (continued)

	Na^+	Ca^{++}	Mg^{++}	Cl^-
AR Lag1	0.9873	0.70071	0.84247	0.955
T	17.4556	16.7293	14.524	16.483
Prob > |T|	<.0001	<.0001	<.0001	<.0001
AR Lag2	−0.2285	.	−0.1425	−0.2356
T	−4.0145	.	−2.455	−4.0563
Prob > |T|	<.0001	.	0.014	<.0001
MA1	.	.	.	0.9954
T	.	.	.	23.236
Prob > |T|	.	.	.	<.0001
SAR Lag12	.	.	.	0.1499
T	.	.	.	2.53
Prob > |T|	.	.	.	0.0117
P-value	0.99	0.99	0.99	0.95
RV	0.2704	1.297	0.081	10353
ME	E	E	E	E

LSW Level smoothing weight, *SL* Smoothed level, *AR* Autoregressive, *SAR* Seasonal autoregressive, *MA* Moving average, *RV* Residual variance (sigma squared), *ME* Model evaluation, *E* Excellent, *SBC* Schwartz Bayesian information criteria, *AIC* Akaike information criteria, Cl^- Chloride ion, HCO_3^- Biocarbonate ion, SO_4^{--} Sulfate ion; Ca^{++} Calcium ion; Mg^{++} = Magnesium ion, Na^+ Sodium ion, *TDS* Total dissolved solids, *Std. Error* Standard deviation

Table 4.6 Characteristics of developed models for Zir-e-Pol station in out let of the Latian Dam

	Na^+	Ca^{++}	Mg^{++}	Cl^-
Suggested Model	ARIMA(2,0,0)(1,0,0)	ARIMA(2,0,0)(0,0,0)	ARIMA(2,0,0)(1,0,0)	ARIMA(2,0,0)(0,0,0)
Equation	$Z_t = 0.566 + (0.728(Z_{t-1}) - 0.151(Z_{t-2}) + a_t) \times (0.1407(Z_{t-12}) + \varepsilon_t)$	$Z_t = 2.186 + 1.02(Z_{t-1}) - 0.338(Z_{t-2}) + a_t$	$Z_t = 1.035 + (0.9301(Z_{t-1}) - 0.309(Z_{t-2}) + a_t) \times (0.1251(Z_{t-12}) + \varepsilon_t)$	$Z_t = 0.491 + 1.1(Z_{t-1}) - 0.3931(Z_{t-2}) + a_t$
AIC	−178.19	193.26	139.59	−247.83
SBC	−163.46	204.38	154.32	−236.78
R-Square	0.6	0.95	0.72	0.9
Std. Error	0.1772	0.3316	0.3042	0.1575
Intercept	0.5663	2.1862	1.035	0.4909
T	20.12	36.474	19.52	15.534
Prob > ITI	<.0001	<.0001	<.0001	<.0001
AR Lag1	0.7281	1.02	0.9301	1.105
T	12.484	18.697	16.682	20.558
Prob > ITI	<.0001	<.0001	<.0001	<.0001
AR Lag2	−0.1513	−0.3385	−0.3091	−0.3931
T	−2.605	−6.1837	−5.543	−7.317
Prob > ITI	0.009	<.0001	<.0001	<.0001
MA1	–	.	.	.
T	–	.	.	.
Prob > ITI	–	.	.	.
SAR Lag12	0.14079	.	0.1251	.
T	2.399	.	2.1409	.

(continued)

4.17 Application of Time Series

Table 4.6 (continued)

	Na^+	Ca^{++}	Mg^{++}	Cl^-
Prob > |T|	0.017	.	0.03	.
P-Value	0.98	0.99	0.97	0.99
RV	0.0314	0.1099	0.0925	0.0248
ME	E	E	E	E
	HCO_3^-	SO_4^{--}	pH	TDS
Suggested Model	ARIMA(1,0,1)(0,0,0)	ARIMA(2,0,0)(1,0,0)	ARIMA(2,0,0)(0,0,0)	ARIMA(1,0,1)(0,0,0)
Equation	$Z_t = 2.45 + 0.628(Z_{t-1}) + a_t + 0.584(a_{t-1})$	$Z_t = 0.88 + (0.997(Z_{t-1}) - 0.381(Z_{t-2}) + a_t) \times (0.2076(Z_{t-12}) + \varepsilon_t)$	$Z_t = 7.775 + 0.842(Z_{t-1}) - 0.142(Z_{t-2}) + a_t$	$Z_t = 228.8 + 0.8(Z_{t-1}) + a_t + 0.591(a_{t-1})$
AIC	141.02	−98.25	63.13	2826.91
SBC	152.07	−83.51	74.24	2837.96
R-Square	0.94	0.53	0.83	0.91
Std. Error	0.305	0.2028	0.267	29.366
Intercept	2.455	0.8803	7.799	228.79
T	32.611	22.965	181.058	16.904
Prob > |T|	<.0001	<.0001	<.0001	<.0001
AR Lag1	0.6279	0.9967	0.7692	0.801
T	12.273	18.302	13.371	21.822
Prob > |T|	<.0001	<.0001	<.0001	<.0001
AR Lag2	–	−0.3812	−0.1255	–
T	–	−6.935	−2.183	–
Prob > |T|	–	<.0001	0.0297	–
MA1	−0.584	.	–	−0.591
T	−10.904	.	–	−11.847

(continued)

Table 4.6 (continued)

	Na$^+$	Ca^{++}	Mg^{++}	Cl$^-$
Prob > |T|	<.0001	.	–	<.0001
SAR Lag12	–	0.2076	–	–
T	–	3.57	–	–
Prob > |T|	–	<.0001	–	–
P-Value	0.99	0.99	0.97	0.99
RV	0.0931	0.0411	0.0713	862.4
ME	E	E	E	E

LSW Level smoothing weight, *SL* Smoothed level, *AR* Autoregressive, *SAR* Seasonal autoregressive, *MA* Moving average, *RV* Residual variance (sigma squared), *ME* Model evaluation, *E* Excellent, *SBC* Schwartz Bayesian information criteria, *AIC* Akaike information criteria, *Cl$^-$* Chloride ion, *HCO$_3^-$* Biocarbonate ion, *SO$_4^{--}$* Sulfate ion, *Ca^{++}* Calcium ion, *Mg^{++}* Magnesium ion, *Na$^+$* Sodium ion, *TDS* Total dissolved solids, *Std. Error* Standard deviation

4.18 Summary

The study indicates Box-Jenkins and exponential time series may be suitable for prediction of water quality.

The developed models for Mg^+, Na^+, SO_4^{-2} parameters in Zir-e-pol station (only outlet), a TDS model in Lavark station and HCO_3^- model in Roodak station described seasonality behaviors and the rest of the models are non- seasonal.

Approximation of the trend of observations shows that the amounts of TDS, Mg^+, Na^+, and SO_4^{-2} parameters are maximum in April and minimum in September. This may be due to a maximum amount of rainfall in early spring and diminishing rainfall in summer.

All the developed models have p-Values above 0.9 which implies they are excellent according to the definition. Comparison of predicted and observation data for the last 6 months shows good conformity. Hence, the developed models are proper and confident and may be useful tools for water quality management in inlets and outlet of water in the dam.

References

Asadollahfardi G (2002) Analysis of surface water quality in Tehran. Water Qual Res J Can 37 (2):489–511

Asadollahfardi G, Rahbar M, Atemiaghda M (2012) Application of time series models to perdict water quality of upstream and downstream of Latian Dam. Universal J of Environ Res Technol 2(1):26–35

Asadollahfardi G, Kodadadi A, Paykani B, Samady Y, Asadollahfardi R (2012) Application of multivariate statistical analysis to define water quality in Jajrud river. Asian J Water Environ Pollut 9(4):1–10

Bendat JS, Piersol AG (1966) Measurement and analysis of random data, John and Wiley, New York

Bowerman BL, O'Connell RT (1987) Time series forecasting. Duxbury Press, Boston, USA

Box GEP, Cox DR (1964) An analysis of transformations. J Roy. Stat. Soc. B 26(2):211–252

Box GEP, Jenkins GM (1976) Time series analysis, forecasting and control. Holden-Day, San Francisco, CA, USA

Caissie D, El-Jabi N, St-Hilaire A (1998) Stochastic modelling of water temperatures in a small stream using air to water relations. Can J Civ Eng 25(2):250–260

Carlson RF, MacCormick A, Watts DG (1970) Application of linear random models to four annual streamflow series. Water Resour Res 6(4):1070–1078

Chatfield C (2002) Time-series forecasting. Chapman and Hall/CRC, Boca Raton

Chow WM (1965) Adaptive control of the exponential smoothing constant. J Ind Eng 16 (5):314–317

Cruz S-L, Yevjevich V (1972) Stochastic structure of water use time series. Colorado University, hydrology paper No.52

Haan CT (1977) Statistical methods in hydrology. The Iowa University Press, Iowa

Hasmida H (2009) Water quality trend at the upper part of johor river in relation to rainfall and runoff pattern. MS thesis, Faculty of Civil Engineeing, Universitiy eknologi, Malaysia

Huck PM, Farquhar GJ (1974) Water quality models using the Box-Jenkins method. J Environ Eng Div 100(3):733–753

Jayawardena A, Lai F (1991) Water quality forecasting using an adaptive ARMA modelling approach. In: Proceedings of the international symposium on environmental hydraulics, pp 1121–1127

Jayawardena AW, Lai FZ (1989) Time series analysis of water quality in Pearl River, China. J Environ Eng ASCE115(3):590–607

Kakavand R (2001) A study of relationship between precipitation and flood hydrograph by statistic and synoptic. MSc Thesis, Tarbiat Moalem University, Tehran, Iran

Kisiel CC (1969) Time series analysis of hydrologic data. Advances in hydro-science, vol 5. Academic press, New York, pp 1–119

Kurunc A, Yurekli K, Cevik O (2004) Performance of two stochastic approaches for forecasting water quality and streamflow data from Yeşilurmak River, Turkey. J Env Model Softw 20:1195–1200

Lohani BN, Wang MM (1987) Water quality data analysis in Chung Kang River. J Environ Eng 113(1):186–195

MacLeod C, Whitfield PH (1996) Seasonal and long-term variations in water quality of the Columbia River at Revelstoke, British Columbia. Northwest Sci 70:55–65

Mahloch JL (1974) Multivariate techniques for water-quality analysis. Environ Eng Div Am Soc Civil Eng 100(EE5):119–1132

Matalas NC (1966) Some aspects of time-series analysis in hydro-logic studies, in statistical method in hydrology. In: Proceeding on hydrology symposium No.5. McGill University, National Research Council of Canada, pp. 271–309

McKenzie (ed) (1984) General exponential smoothing and the equivalent ARMA process. J Forecast 3(3):333–344

McMichael FC, Hunter JS (1972) Stochastic modeling of temperature and flow in rivers. Water Resour Res 8(1):87–98

Thomann RV (1967) Time series analysis of water quality data. J Sanit Eng Div Amer Soc Civil Eng 93:1–23

Winters PR (1960) Forecasting sales by exponentially weighted moving averages. Manage Sci 6:324–342

Chapter 5
Artificial Neural Network

Abstract Often in water quality management, understanding the relationship between input and output data might be a complicated process. In this situation Data Driven Models using information and collected data (input data) find out the relationship between inputs and outputs. In this regard, Artificial Neural Network (ANN) is one of the Data Driven Models which has recently been applied as a tool for modeling complicated processes. In this chapter, after reviewing the developing process of ANN in water quality management, the theory of the ANN is mentioned in detail for both static and dynamic methods. Data preparation, learning rate and model efficiency including selection of number of neurons in hiding layer which has a minimum error in learning rate and network efficiency is described in detail. At the end step, as a case study water quality of Zaribar Lake located in the Northwestern part of Iran, using Multilayer Perceptron (MLP) neural network method are described.

5.1 Introduction

Lack of water resources and optimum management has been two recent challenges of water resources engineering. Population growth, decline of useable water resources, improvements in lifestyle, growing rate of consumption, climate change and several other parameters have caused useable water to be a noteworthy problem for the future. Economic and efficient use of water resources and its management have an increasingly important role.

Other challenges which water quality managements and environmental engineers are facing are controlling the nutrients released to surface waters. Despite all efforts, eutrophication is also other major problems with water quality management. Eutrophication is defined as a cultural or accelerated enrichment of nutrient in lakes, rivers, estuaries and marine waters in which the natural eutrophication process has gone forward by hundreds or more years of human activities that add nutrients (Burkholder 2000). Two important parameters which cause eutrophication are phosphor and nitrogen. Analyzing sample data of Phosphor and Nitrogen parameters, in surface water is necessary for understanding the eutrophication situation and the development of a model to predict understanding would help to manage water resources effectively. There are different methods for data analyses, such as statistic techniques. For Prediction of water quality parameters, first, accurate study of different processes which can affect water quality and developing statistical or deterministic models according to the obtained information is necessary. Second, developing Data Driven Models using information and collected data is an essential tool. In the latter technique, relationship between input and output data can be found using input data, but still physical understanding of the phenomena is significant for having proper input data for model; however, it is not needed to simulate a complicate processes. ANN is one of the Data Driven Models which has recently been applied as a tool for modeling complicated processes.

5.2 Historical Background

French and Recknagel (1994) used the back propagation of ANN to predict seven kinds of algae with nine defined environmental variables in a Reendbatch tank in Germany and obtained acceptable results. Recknagel et al. (1997) developed and validated the ANN by limn logical time series from four different freshwater systems. The water-specific time-series comprised cell numbers or biomass of the ten dominating algae—species as observed over up to 12 years and the measured environmental driving variable. The resulting prediction of the ANN can fit the nonlinearity of ecological phenomena to a high degree. Maier et al. (1998) applied the ANN back propagation type to model a group of Ciano bacteria type of Cyanobacteria Anabaena in Murray River in Morgan, North Australia. The results were relatively successful in providing a good forecast for both incidence and magnitude of growth peak. Wei et al. (2001) applied the ANN to forecast algal blooms and they achieved accurate result. Rectnagel et al. (2002) compared potentials and the achievement of the ANN and genetic algorithms in the term of forecasting and understanding of algal bloom in Lake Kasumigaura, Japan. The ANN predicted the timing and the magnitude of the algal bloom 7 days in advance. Karul et al. (2000) also modeled the eutrophication process of Keban dam, reservoir Morgan and Eymir Lake uses ANN and the results were successful. Wilson and Recknagel (2001) suggested a generalized architecture of a feed forward ANN for prediction of algal abundance, their model was validated by mean of time-series data from six different freshwater lakes. Huang and Foo (2002) studied neural network modeling of salinity variation in Apalachicola River and their results was acceptable. French and Recknagel (1994) developed a model to quantify the interaction between biotic factors and algal genera in Lake Kasumigaura, Japan uses ANN method and results showed that the timing and magnitude of algal blooms of Microcystis phormidium and Synedra in the mentioned lake could be successfully predicted. Markus et al. (2003) applied the ANN to predict uncertainty of weekly nitrate-nitrogen. Panda et al. (2004) used the ANN to predict water quality using satellite imagery data and thus, indicates that it has the potential to make the water quality determination process cost effective, quick and sensible. Jiang et al. (2006)

studied an improved back propagation of the ANN model for eutrophication of eastern china, the model was applied to four eastern lakes and results described the ANN is suitable for predicting eutrophication. Kuo et al. (2007) used back-propagation neural network to relate the number of water quality indicators such as DO, TP, Chl-a and secchi dish depth in a reservoir in central Taiwan and they concluded that the ANN is able to predict these indicators with reasonable accuracy. Kanani et al. (2008) predicted the salinity levels for 1 month in advance in a Talkheh Rud River (Iran) by applying the MLP and IDNN models, the result was reasonable The Illinois state water survey conducted a study to assess the potential of the ANN in forecasting weekly nitrate-nitrogen concentration. Three ANN models were applied to predict weekly Nitrate—N concentration in the Sangamon River near Decatar, Illinois, based on the previous week's precipitation, air temperature, and discharge and past nitrate concentration. The result model was more accurate than the linear regression model having the same input and output (Momcilo et al. 2003). Asadollahfardi et al. (2010) applied the MLP for predication of eutrophication in Anzalily Wetland using TN, TP as input in the MLP and predicted BOD parameters, the results indicated reasonable accuracy. Asadollahfardi et al. (2011) studied static and dynamic neural network to TDS of Talkheh-Rud River, Iran, and predicted the future of salinity. Asadollahfardi et al. (2013) predicted sodium adsorbtion ratio (SAR) of Chelghazy River in Kurdistan (Iran) using MLP neural network, and the results was acceptable.

5.3 Artificial Neural Network Theory

5.3.1 Theory of ANN

Bearing in mind natural neural and its components, scientists developed an artificial neural system. This is the smallest unit of an ANN. An artificial neural system consists of three components including weighting (W), bias (b) and transfer function (f). These three components are unique to each neural system. In Fig. 5.1, "p" and "n" are input and output while "a" is net output. The junctions 1 and 2 in the figure indicate the schematic of an artificial neural system. Function of an artificial neural network would be called "p".

$$n = wp + b \tag{5.1}$$

$$a = f(n) = f(wp + b) \tag{5.2}$$

Generally, the ANN is divided into two groups, static and dynamic. Time is not a key parameter in the static of the ANN but it is one of the main parameters in dynamic networks.

Hornik et al. (1989) proved the "universal approximator theory" which expressed that a feedforward neural network with a hidden layer of sigmoid tangent and linear output layer would be able to estimate each complex function (Cybenko 1989;

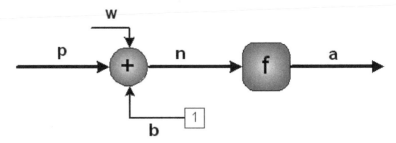

Fig. 5.1 Schematic of an artificial neuron

Hornik 1991, 1993; Leshno et al. 1993). Figure 5.2 presents the schematic of the ANN. Figure 5.3 shows schematic tangent—sigmoid transfer function and linear transfer function. Number of neurons in hidden layers for each model may be obtained using trial and error. In this network, number of input vector component.

R = the number of input vector components; S_1 and S_2 = number of Neurals in hidden and output layers, respectively.

Number of network output would be S_2. Function of the network can be modeled by Eqs. 5.3 and 5.4 (Menhaj 1998):

$$a_j^1(t) = F\left\{\sum_{i=1}^{R} w_{j,i}^1 p_i(t) + b_j^1\right\} \qquad (5.3)$$

$$a_j^2(t) = G\left\{\sum_{j=1}^{s1} w_{k,i}^2 a_J^1(t) + b_k^2\right\} \qquad (5.4)$$

Fig. 5.2 Transfer function tangent sigmoid

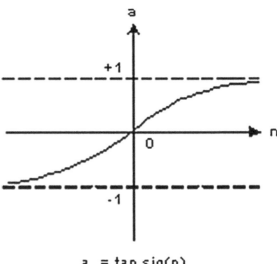

5.3 Artificial Neural Network Theory

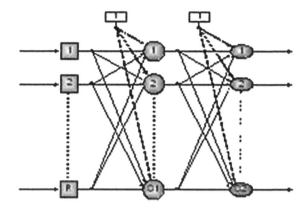

Fig. 5.3 Schematic of MLP with a hidden layer

where R = numbers of input vector components. S_1 and S_2 = numbers of neurons in hidden and output layers, respectively. P = input vector. w^1 and w^2 = weighting matrix in hidden and output layers, respectively. $b^1 = b^2$ are bias vectors in Hidden and output layers, respectively. G and F = neuron transfer functions in hidden and output layers respectively (Menhaj and Safepoor 1998).

For further reading see references (Menhaj and Safepoor 1998).

5.3.2 Dynamic ANN Models

The inputs of dynamic are the same as the static model; the difference is that the effect of the past period is considered in this model. There are several methods by which a static model can be turned to dynamic. One of them is Time Delay Neural Network (TDNN) operators. A TDNN operator receives an input signal and keeps it for a time step. And in the next time step the input signal emerges as an output result. By connecting N series of TDNN operator, Tapped Delay Line (TDL) will be obtained. The output is a vector with N + 1 components. The N + 1 components include the input in the current time step and N time steps before.

5.3.3 Data Preparation

Considering the application of sigmoid tangent in the hidden layers of the networks and the special formula of this function, the scale of input data have to be changed. Considering Fig. 5.2 sigmoid tangent function, it is clear that the slope of this function is differentiated according to the sums in the interval (1, −1) of the ambient and has few changes out of this interval. For all data, output and input should be transformed to the (1, −1) interval to prevent the network saturation (Asadollahfardi 2012).

5.3.4 Learning Rate

There is a parameter called learning rate in the training algorithm of back propagation, which is on the basis of the steepest descent. Its aim is to minimize the sum square error of outputs. Determining the proper learning rate is one of the most sensitive processes to use the algorithm of back propagation (Menhaj and Safepoor 1998). The learning rate is indicated by a symbol α and determines the velocity of convergence in this algorithm. The performance of the steepest descent algorithm is improved if the learning rate is permitted to change during the training process. An adaptive learning rate attempts to make the learning step as big as possible to keep the learning stable and requires some alters in the training procedure (Asadollahfardi 2012).

5.3.5 Model Efficiency

Three error criterions of VE, MAE, and RMSE are used to evaluate the output of obtained models. Equations (5.5), (5.6) and (5.7) show these expressions (Kennedy and Neville 1976):

$$\text{Volume Error (VE)} = \frac{1}{T}\sum_{t=1}^{T} \left| \frac{Obs_t - For_t}{Obs_t} \right| \times 100 \qquad (5.5)$$

$$\text{Mean Absolute Error (MAE)} = \frac{1}{T}\sum_{t=1}^{T} |Obs_t - For_t| \qquad (5.6)$$

$$\text{Root Mean Square Error (RMSE)} = \sqrt{\frac{1}{T}\sum_{t=1}^{T}(Obs_t - For_t)^2} \qquad (5.7)$$

where, T = discrete time, t = length of time series, Obs_t = observed parameter in time of t ($1 \leq t \leq T$) and For_t = predicted parameter in time of t ($1 \leq t \leq T$). Also the correlation coefficient R is applied to show the validity between real data and predicted ones which are described in Eq. 5.6.

$$R = \frac{\sum(x - \bar{x})(y - \bar{y})}{\sqrt{\sum(x - \bar{x})^2 \sum(y - \bar{y})^2}} \qquad (5.8)$$

where $\bar{x}, \bar{y}$ = means of x and y series. R shows the relationship between observed data and predicted data. If relations are very strong, R approaches one.

5.4 Application of Artificial Neural Network

5.4.1 A Case Study: Zaribar Water Quality (Iran)

Zaribar Lake is located near the North West of Marivan (North West of Iran), in 46°, 6' longitude and 35°, 31' latitude and at an altitude of 1,250 m from the sea surface. The lake area varies due to the variation in water volume during different seasons of the year. The minimum depth is about 2 m and the maximum is 6 m. The perimeter of the lake is about 22.5 km. The average rate of rainfall is 786 mm per year and relative humidity is 58.4 %; average vaporization is 1,900 mm annually. Figure 5.4 and Table 5.1 indicates the location and a statistical summary of water quality data in Zaribar Lake, respectively (Irani 1991).

Sewages from Marivan city and its surrounding villages are discharged into Zaribar Lake. This resulted in high amounts of Phosphor and Nitrogen and phosphate and nitrate chemicals which are washed up by rain water from the adjacent farmland. Unfortunately, this causes a decrease of DO of water by growth of algae. This may cause an eutrifcation phenomena in the lake. Figure 5.4 presents the situation of Mari van (Zaribar Lake) in the map of Iran.

The aim of this case study was to find effects of TP and TN parameters in increasing and predicating BOD parameter (eutrophication) in Zaribar Lake using the MLP technique and indicating application of MLP neural network.

Twelve year data of Zaribar Lake (Table 5.1) with 1 month time delay was used to the modeling process. Eight years of the data were used for network training and

Fig. 5.4 Situation of Marivan (Zaribar Lake) in map of Iran

Table 5.1 Statistic summary of the data of the Zaribar lake from 1949 to 2006

Parameters	No.	Minimum	Maximum	Standard deviation	Mean	Median	10 percentile	25 percentile	75 percentile	Skewness
TN (mg/l)	144	0.56	7.99	1.15	4.21	4.14	2.83	3.49	4.77	0.43
TP (mg/l)	144	0.1	1.6	0.41	0.63	0.5	0.2	0.25	1.03	0.42
BOD (mg/l)	144	13.2	109.2	20.39	38.35	33.6	15.96	22.8	48	1.22

5.4 Application of Artificial Neural Network

Table 5.2 Variation of errors in hidden layer when applying TP data as an input

No. of neurons	VE error—TP input			MAE error—TP input			RMSE error—TP input		
	Training	Test	Total	Training	Test	Total	Training	Test	Total
2	26.365	26.419	26.379	19.248	19.369	19.284	19.232	18.432	18.291
3	24.198	24.212	24.202	18.017	18.118	29.098	18.965	17.543	17.408
4	27.110	27.164	27.124	19.920	20.046	32.617	20.667	19.240	19.093
5	36.702	36.718	36.706	25.495	25.631	44.073	24.899	23.430	23.248
6	26.397	26.451	26.411	19.370	19.520	31.821	20.093	18.710	18.528
7	30.598	30.655	30.613	21.775	21.886	36.858	21.883	20.370	20.244
8	25.301	25.355	25.315	18.605	18.799	30.570	19.363	18.070	17.825
9	21.047	21.098	21.060	15.360	15.769	25.501	16.000	15.120	14.574
10	28.492	28.548	28.507	18.236	18.516	34.348	16.534	15.170	14.806
11	22.111	22.163	22.124	14.134	14.553	26.756	13.003	11.910	11.350
12	15.729	15.777	15.741	9.827	9.920	19.146	9.108	7.310	7.205
13	8.310	8.355	8.321	5.556	5.833	10.335	5.900	4.560	4.196
14	14.665	14.713	14.677	10.028	10.261	17.972	10.132	8.770	8.469
15	18.920	18.970	18.933	13.788	13.941	23.123	14.427	13.020	12.831
16	29.556	29.612	29.570	20.959	21.190	35.875	21.032	19.740	19.446
17	33.811	33.869	33.826	24.844	25.470	41.040	25.632	25.190	24.343
18	37.002	37.061	37.017	28.050	28.757	44.929	29.662	29.430	28.471
19	39.129	39.190	39.145	29.352	30.184	47.500	30.728	30.730	29.596
20	42.320	42.382	42.336	31.108	31.691	51.325	31.940	31.410	30.626

Table 5.3 Variation of errors in hidden layer when applying TN data as an input

No. of neurons	VE error—TN input			MAE error—TN input			RMSE error—TN input		
	Training	Test	Total	Training	Test	Total	Training	Test	Total
2	39.742	40.160	39.867	26.776	27.058	26.861	28.533	28.833	28.623
3	37.923	38.321	38.043	25.551	23.273	25.631	27.227	27.513	27.313
4	41.395	41.830	41.526	27.890	25.404	27.978	29.720	30.032	29.813
5	49.951	50.489	50.113	33.655	30.663	33.764	35.863	36.248	35.978
6	40.197	40.734	40.358	27.083	24.739	27.191	28.859	29.245	28.975
7	43.793	44.165	43.904	29.505	26.822	29.581	31.441	31.708	31.521
8	38.688	39.411	38.905	26.066	23.935	26.212	27.776	28.295	27.932
9	31.703	33.315	32.186	21.360	20.233	21.686	22.761	23.918	23.108
10	32.343	33.418	32.666	21.791	20.295	22.009	23.221	23.992	23.452
11	25.027	26.681	25.523	16.862	16.204	17.196	17.968	19.155	18.324
12	16.864	17.174	16.957	11.362	10.430	11.425	12.107	12.330	12.174
13	10.416	11.491	10.738	7.018	6.979	7.235	7.478	8.250	7.710
14	19.303	20.191	19.569	13.005	12.263	13.185	13.858	14.496	14.050
15	28.417	28.975	28.584	19.146	17.597	19.259	20.402	20.802	20.522
16	41.995	42.863	42.255	28.294	26.031	28.470	30.150	30.773	30.337
17	51.625	54.126	52.376	34.783	32.872	35.288	37.064	38.860	37.603
18	60.057	62.889	60.907	40.464	38.194	41.036	43.118	45.151	43.728
19	62.227	65.575	63.232	41.926	39.825	42.603	44.676	47.080	45.397
20	64.666	66.981	65.360	43.569	40.679	44.037	46.427	48.089	46.925

5.4 Application of Artificial Neural Network

remaining data for the testing of the network. The software applied for modeling was Neural Network Toolbox and was the MATLAB version (2007).

Data used, including the amount of TN and TP and BOD parameters was monitored by the Water Authority of Kurdistan province (Iran) for a period of 12 years from 1994 to 2006. This was the only data available for this work.

Considering the theory of universal approximation as mentioned previously, all of the ANN which was applied in this study has relatively similar structure with the main difference being the numbers of neurons in the hidden layer.

Tables 5.2 and 5.3 present the amount of errors in training, testing and total with various number of neuron using VE, MAE, RMSE standard errors when input data were TP and TN in case of using 2–20 neurons in the hidden layer. As indicated in the tables, using both parameters of TP and TN, minimum errors obtained when 13 neurons were applied in the hidden layer. Correlation of determination for TP input into training, testing and total were 0.981, 0.964 and 0.974 respectively, which indicates reasonable conformity between actual BOD parameters data and the model predictions (Figs. 5.5, 5.6 and 5.7) and when TN is applied as input to the model, the correlation of determination in training, testing and total are 0.963, 0.957 and 0.968 respectively (Figs. 5.8, 5.9 and 5.10). In Figs. 5.11 and 5.12 the conformity between predicted BOD parameter and real one when input parameters were TP and TN is illustrated.

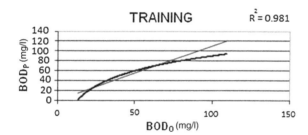

Fig. 5.5 Correlation of determination diagram between actual BOD (mg/l) data and prediction (TN as an input data)

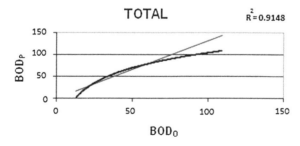

Fig. 5.6 Correlation of determination diagram between actual BOD (mg/l) data and prediction (TN as the input data)

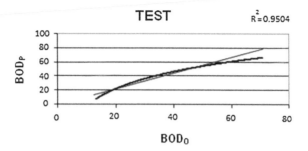

Fig. 5.7 Correlation of determination diagram between actual BOD (mg/l) data and prediction (TN as the input data)

Fig. 5.8 Correlation of determination diagram between actual BOD (mg/l) data and prediction (TP as the input data)

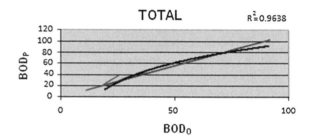

Fig. 5.9 Correlation of determination diagram between actual BOD (mg/l) data and prediction (TP as input data)

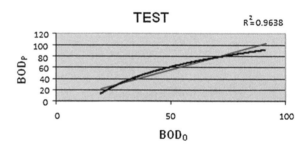

Fig. 5.10 Correlation of determination diagram between actual BOD (mg/l) data and prediction (TP as the input data)

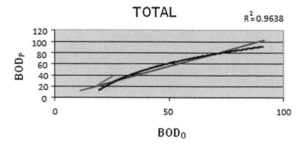

5.4 Application of Artificial Neural Network

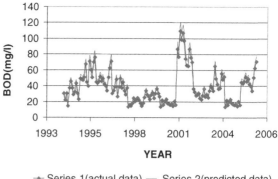

Fig. 5.11 Comparison between BOD prediction and actual BOD data, considering TN as input data

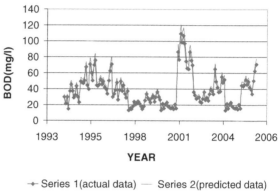

Fig. 5.12 Comparison between BOD prediction and actual BOD data, considering TP as input data

TP parameter was found to have marginally less error than TN when used as an input for the MLP model. This may present that TP parameter may have a greater role to play than TN in generating pollution (BOD) in the wetland. This can be proved by applying sensitively analyze.

5.4.1.1 Comment

As a general, MLP neural network is a suitable tool for analyzing surface water quality and comment related to mentioned case study is as follows:

In the testing model, with 13 neurons in the hidden layer, when TP was used as an input parameter to the MLP model for prediction of the BOD, the error was 8.32 % which was less than the error obtained with TN as the input. This may describe that TP is more influential in producing pollution (BOD) than TN. This was also the minimum error in this study. Also, the correlation coefficients in training, testing and total were 0.981, 0.964 and 0.974 respectively, which shows reasonable conformity between real BOD parameter data and the model predictions.

The least amount error when applying the TN as an input to the MLP model was 10.73 % with 13 neuron chosen in the hidden layer. Correlation coefficients in training, testing and total are 0.963, 0.957 and 0.968 respectively, which may indicate acceptability of the MLP model and may present the role of the TN parameter in contamination of the wetland.

As a whole, the results of this study may help to decision makers in water quality management in Zaribar Lake and application of the technique may apply to another part of the world.

References

Asadollahfardi G, Khodadadi A, Gharayloo R (2010) The assessment of effective factors on Anzali wetland pollution using artificial neural networks. Asian J Water Environ Pollut 7(2):23–30

Asadollahfardi G, Hemati A, Moradinejad S, Asadollahfardi R (2013) Sodium adsorption ratio (SAR) prediction of the Chalghazi river using artificial neural network (ANN) Iran. Curr World Environ 8(2):169–178

Asadollahfardi G, Taklify A, Ghanbari A (2012) Application of artificial neural network to predict TDS in Talkheh Rud River. J Irrig Drain Eng 138(4):363–370

Burkholder J (2000) Critical needs in harmful algal bloom research. Opportunities for environmental applications of marine biotechnology national academy of sciences. National Research Council, Washington, DC, pp 126–149

Cybenko G (1989) Approximation by superpositions of a sigmoidal function. Math Control Signals Syst 2(4):303–314

French M, Recknagel F (1994) Modeling of algal blooms in fresh water using artificial neural network. Computational Mechanic INC, Billerica, MA01821 (USA), 87–94. Hydrol Eng ASCE 5(2):123–137

Hornik K (1991) Approximation capabilities of multilayer feedforward networks. Neural Netw 4(2):251–257

Hornik K (1993) Some new results on neural network approximation. Neural Netw 6(8):1069–1072

Hornik K, StInchcombe M, White H (1989) Forward networks are universal approximators. Neural Netw 2(5):359–366

Huang W, Foo S (2002) Neural network modeling of salinity variation in Apalachicola River. Water Res 36(1):356–362

Irani J (1991) Study of hydro climatology of basin of Zarivar (Marivan), MSc thesis, Earth Science

Jiang Y, Xu Z, Yin H (2006) Study on improved BP artificial neural networks in eutrophication assessment of China eastern lakes. J Hydrodyn Ser B 18(3):528–532

Kanani S, Asadollahfardi G, Ghanbari A (2008) Application of artificial neural network to predict total dissolved solid in Achechay River basin. World Appl Sci J 4(5):646–654

Karul C, Soyupak S, Çilesiz AF, Akbay N, Germen E (2000) Case studies on the use of neural networks in eutrophication modeling. Ecol Model 134(2):145–152

Kennedy J, Neville A (1976) Basic statistical methods for engineers and scientists. Harper and Row, New York

Kuo J-T, Hsieh M-H, Lung W-S, She N (2007) Using artificial neural network for reservoir eutrophication prediction. Ecol Model 200(1):171–177

Leshno M, Lin VY, Pinkus A, Schocken S (1993) Multilayer feedforward networks with a nonpolynomial activation function can approximate any function. Neural Netw 6(6):861–867

References

Maier HR, Dandy GC, Burch MD (1998) Use of artificial neural networks for modelling cyanobacteria Anabaena spp. in the River Murray, South Australia. Ecol Model 105(2):257–272

Markus M, Tsai CW-S, Demissie M (2003) Uncertainty of weekly nitrate-nitrogen forecasts using artificial neural networks. J Environ Eng 129(3):267–274

MATLAB Version 7.0 (2007) Mathworks, Inc., Natick, MA

Menhaj M (1998) Artificial calculating, principal of ANN, vol 1. Amirkaber University Publisher, Iran (In Persian)

Menhaj M, Safepoor N (1998) Artificial calculating, vol 2. Professor Hesabi Publisher, Iran (In Persian)

Momcilo M, Christina WS, Tsai, Misganaw D (2003) Uncertainty of weekly nitrate-nitrogen forecasts using artificial neural networks. J Environ Eng 129:267–274

Panda S, Garg V, Chaubey I (2004) Artificial neural networks application in lake water quality estimation using satellite imagery. J Environ Inform 4(2):65–74

Recknagel F, Bobbin J, Whigham P, Wilson H (2002) Comparative application of artificial neural networks and genetic algorithms for multivariate time-series modelling of algal blooms in freshwater lakes. J Hydroinform 4:125–133

Recknagel F, French M, Harkonen P, Yabunaka K-I (1997) Artificial neural network approach for modelling and prediction of algal blooms. Ecol Model 96(1):11–28

Wei B, Sugiura N, Maekawa T (2001) Use of artificial neural network in the prediction of algal blooms. Water Res 35(8):2022–2028

Wilson H, Recknagel F (2001) Towards a generic artificial neural network model for dynamic predictions of algal abundance in freshwater lakes. Ecol Model 146(1):69–84

Chapter 6
Introducing of Ce-Qual-W2 Model and Its Application

Abstract Water authority organizations are interested in information the existing situation, seasonal variations and expectations of the future situation of the water quality parameters of surface and groundwater. While detail information about surface water or ground water are available, deterministic models for predicting future values of water quality is more proper than stochastic models. In this regard, numerical models demonstrated an impressive capacity to support important water resource decisions. Therefore, in this chapter Ce-Quel-W2 and Qual-2K models as numerical models that are applied to simulate water quality are described in details. At the end, using Ce-Qual-W2 model the water quality of Karkheh Dam, which located in the Northwestern province of Khūzestān in Iran as a case study, is investigated. In addition the application of Qual-2K models for simulating water quality of the Kārūn River, which located in the south west of Iran, was described.

6.1 Introduction

The first step of a water quality study is to identify the relevant parameter, which affects the water quality. Rivers and streams are an important component of the natural environment, and need to be protected from all sources of pollution because man's own survival depends on their use. Rivers; however, are increasingly under human threat from different pollutants, which include conventional pollutants (organic matter and inorganic nutrients) and hazardous substances (organic contaminants and heavy metals). Despite the fact that the river water quality can be influenced by natural phenomena such as climate and geology (Boorman 2003), the main sources of pollution are related to anthropogenic activities: mining, agriculture, forestry, cattle farming and urbanization. As a result, river water quality is affected by both point and diffuse sources of pollution. To tackle these typical water quality problems, and for the sake of both ecological and human welfare, rivers (all water resources in general) must be protected, restored and sustained (Deksissa 2004).

There are many water quality parameters, but TDS is an important water quality parameter especially in reservoirs. In fact, TDS can be considered a pollutant for this reason, it is vital to have information about the existing situation, seasonal variations and expectations of the future situation of the parameter.

The Department of Environment (DOE) and Institute of Standards and Industrial Research of Iran (ISIRI) (2007) have specified that the permitted amount of TDS in drinking water should be less than 500 mg/l (milligrams per liter) and water for irrigation should be 1,000 mg/l.

If the amount of TDS increases above the standard level, it may cause some problems which are as follows (Tebbutt 1997):

1. Providing an undesirable taste in water.
2. If there is magnesium or calcium carbonate in the water, they may cause a core problem in the equipment of a dam such as erosion.
3. Increasing cost of water treatment to reduce the amount of TDS.
4. The increasing amount of TDS in water causes a reduction in dissolved oxygen.

In practice, a minimum in stream flow standards are usually based on some combination of the following: historic discharge, channel morphology, water quality, the ecology of aquatic species, empirical evidence, modeling, and ultimately arbitrated between user groups. For a particular area, in stream flow requirements will depend on local and downstream conditions and can vary considerably within areas of similar climate and hydrology (Beecher 1990).

To predict the damage caused by these problems, modeling of the water resources is essential. The need for predictive water quality modeling has arisen largely as a result of increased eutrophication of lakes throughout the world (Canfield and Bachmann 1981).

Numerical models have demonstrated an impressive capacity to support important water resource decisions. Models are typically used to support development and public policy decisions in a variety of areas: simulation of discharges, outfalls, and intakes; changes to wastewater treatment systems; approval of changes in industrial processes; operation of dams and reservoirs; and water resource allocations, among other uses. The value of modeling is important in economic and financial terms with regard to determining particular project options and phased investment programs (Cox 2003).

6.2 Historical Background

There are many studies for modeling river and reservoir water quality which were used Ce-Qual-W2 model.

Previous studies used the Dyresm model for Mymeh dam in Iran, using salinity and temperature data from 1970 to 1979. The result indicated, there were very weak stratification (TDS) during winter and stratification started in spring and during the summer there was strong stratification. Comparison between salinity before and after construction of the dam across the river described a reduction in salinity (Shiati 1996).

Bookan dam in Kurdistan province across the Zareneh River in Iran was also studied for variations in salinity. The result of the study presented weak stratification (TDS) (Sarang and Tajreshi 2001). This may be due to the low height of the reservoir and the temperate climate situation.

TDS and temperature parameters were also evaluated in long lake reservoir with 40 km length across Spokane River and compared to the data generated using the Ce-Qual-W2 model for simulation. The result of the model and the observed data were the same (Berger and Wells 2001).

Using the Dyresm model to evaluate TDS data between 1961–1990 in Raes Ali Delvary above the Shapoor dam in Iran, the result indicated strong stratification of TDS and Temperature.

Rounds (2001) indicated success and failure of CEQUAL-W2 by a modeling study of the Tualatin River in northwestern Oregon. In his study, CEQUAL-W2 was used successfully to assess the sources and transport of phosphorus, quantify the river's ammonia assimilative capacity, determine the relative significance of the sources and sinks of dissolved oxygen, quantify the factors that affect phytoplankton growth, and test the effects of potential management strategies.

Razdar et al. (2011) compared the results of the CE-QUAL-W2 model with the WASP5 and MIKE11 models to assess the water quality of Pasikhan River. The contaminant loadings of Nitrate and Phosphate was utilized in the CE-QUAL-W2, WASP5 and MIKE11 simulations. The sensitivity analysis for CE-QUAL-W2 model indicated that the model is highly sensitive to the Manning coefficient and point source flow rate. The calibrated model responses was in good agreement with the actual data and could be applied as scenario generators in a general strategy to conserveor improve the water quality. During the period of intense stratification, forecasting from CE-QUALW2 are inconsistent better to the actual data than those from Mike11 and WASP5 due to the improved transport scheme applied in CE-QUAL-W2.

6.3 Theory of the Model

Generally the study of water quality in a reservoir consists of two sections which are as follows:

1. Experimental study
2. Water quality simulation of mathematical model

Domain equations in the Ce-Qual-W2 program are presented in Table 6.1, where, in this chapter the QUAL2K (Chapra and Pelletier 2003) stream and river quality model was used. QUAL2K is a modernized version of the QUAL2E (or Q2E) model (Brown and Barnwell 1987). The model is supplied by the Watershed &

Table 6.1 Governing equations in Ce-Qual-W2 model

Equation	The governing equation assuming no channel slope	The governing equation assuming an arbitrary channel slope and conservation of momentum at branch intersections
X-momentum	$\dfrac{\partial UB}{\partial t}+\dfrac{\partial UUB}{\partial x}+\dfrac{\partial WUB}{\partial z}=$ $gB\dfrac{\partial \eta}{\partial x}-\dfrac{gB}{\rho}\int_{\eta}^{z}\dfrac{\partial \rho}{\partial x}dz+$ $\dfrac{1}{\rho}\dfrac{\partial Bt_{xx}}{\partial x}+\dfrac{1}{\rho}\dfrac{\partial Bt_{xz}}{\partial z}$	$\dfrac{\partial UB}{\partial t}+\dfrac{\partial UUB}{\partial x}+\dfrac{\partial WUB}{\partial z}=gB\sin\alpha$ $+g\cos\alpha B\dfrac{\partial \eta}{\partial x}-\dfrac{g\cos\alpha B}{\rho}\int_{\eta}^{z}\dfrac{\partial \rho}{\partial x}dz$ $+\dfrac{1}{\rho}\dfrac{\partial Bt_{xx}}{\partial x}+\dfrac{1}{\rho}\dfrac{\partial Bt_{xz}}{\partial z}+qBU_x$
Z-momentum	$0=g-\dfrac{1}{\rho}\dfrac{\partial P}{\partial z}$	$0=g\cos\alpha-\dfrac{1}{\rho}\dfrac{\partial P}{\partial z}$
Continuity	$\dfrac{\partial UB}{\partial x}+\dfrac{\partial WB}{\partial z}=qB$	$\dfrac{\partial UB}{\partial x}+\dfrac{\partial WB}{\partial z}=qB$
State	$\rho=f(T_w,\phi_{TDS},\phi_{XS})$	$\rho=f(T_w,\phi_{TDS},\phi_{XS})$
Free surface	$B_\eta\dfrac{\partial \eta}{\partial t}=\dfrac{\partial}{\partial x}\int_{\eta}^{h}UBdz-\int_{\eta}^{h}qBdz$	$B_\eta\dfrac{\partial \eta}{\partial t}=\dfrac{\partial}{\partial x}\int_{\eta}^{h}UBdz-\int_{\eta}^{h}qBdz$

U = horizontal velocity, ms^{-4}; W = vertical velocity, ms^{-4}; B = channel width; P = pressure; τ_x = x-direction lateral average shear stress; τ_y = y-direction lateral average shear stress; ρ = density; η = water surface

Water Quality Technical Support Center of the EPA (U.S. Environmental Protection Agency).

The U.S. EPA Enhanced Stream Water Quality Model (QUAL2K) is frequently used to simulate water quality in streams receiving pollutant discharges. The model capabilities that are relevant to water quality include reactions of carbonaceous, nitrogenous, and benthic oxygen demand, atmospheric reaeration, and the effects of these processes on the dissolved oxygen in a receiving stream.

The model, integrates inputs from point and non-point sources to determine impacts on water quality in receiving water bodies. This model determines assimilative capacities of the water body, level of best management practices, or allows predicting the time required for a system to recover after being altered.

QUAL2K simulates up to 15 water quality constituents in branching stream systems. The model uses a finite-difference solution of the advective-dispersive mass transport and reaction equations. A stream reach is divided into a number of computational elements, and for each computational element, a hydrologic balance in terms of stream flow (e.g., m/s), a heat balance in terms of temperature (e.g., 3 °C), and a material balance in terms of concentration (e.g., mg/l) are written.

Both advective and dispersive transport processes are considered in the material balance. Mass is gained or lost from the computational element by transport processes, wastewater discharges, and withdrawals. Mass can also be gained or lost by internal processes such as release of mass from benthic sources or biological transformations.

6.3 Theory of the Model

The program simulates changes in flow conditions along the stream by computing a series of steady state water surface profiles. The calculated stream flow rate, velocity, cross-sectional area, and water depth serve as a basis for determining the heat and mass fluxes into and out of each computational element due to flow. Mass balance determines the concentrations of conservative minerals, coliform bacteria, and non-conservative constituents at each computational element.

In addition to material fluxes, major processes included in the mass balance are the transformation of nutrients, algal production, benthic and carbonaceous demand, atmospheric reaeration, and the effect of these processes on the dissolved oxygen balance. QUAL2K uses chlorophyll a as the indicator of planktonic algae biomass. The nitrogen cycle is divided into four components: organic nitrogen, ammonia nitrogen, nitrite nitrogen, and nitrate nitrogen. In a similar manner, the phosphorus cycle is modeled by using two components. The primary internal sink of dissolved oxygen in the model is biochemical oxygen demand (BOD). The major sources of dissolved oxygen are algal photosynthesis and atmospheric reaeration (Khodadadi Darban 2010) .

The hydrodynamic model was developed on the foundation of the continuity equation, the momentum equation, and the mass-balance equation for salt. The water quality model is based on the laterally integrated equation describing the mass-balance of a dissolved or suspended substance in the water column (Eq. (6.1)).

$$\frac{\partial(CB)}{\partial t} + \frac{\partial(CBu)}{\partial t} + \frac{\partial(CBw)}{\partial t} = \frac{\partial}{\partial x}\left(K_x B \frac{\partial C}{\partial x}\right) + \frac{\partial}{\partial z}\left(K_z B \frac{\partial C}{\partial z}\right) + BS_i + BS_e \quad (6.1)$$

where; t = time [T]; x = distance seaward along river axis [L]; z = distance upward in the vertical direction [L]; B = river width [L]; C = laterally averaged concentration [M/L3]; u and w = laterally averaged velocities in the x and z directions, respectively [L/T]; K_x and K_z = turbulent diffusion coefficients in the x and z directions, respectively [L2/T]; S_i = time rate of internal increase (or decrease) by biochemical reaction processes [M/L3T]; S_e = time rate of external addition (or withdrawal) across the boundaries [M/L3T].

As indicated on Fig. 6.1, the water quality model consists of eight interlinked components including organic nitrogen (ON), ammonium nitrogen (NH_4^-N), nitrite–nitrate nitrogen ($NO_2^+NO_3^-N$), organic phosphorus (OP), inorganic phosphorus (PO_4^-P), chlorophyll 'a' (chl), carbonaceous biochemical oxygen demand (CBOD), and the dissolved oxygen (DO) (Khodadadi Darban 2010).

Each of the water quality components can be represented by the same equation as Eq. (6.1), but with its own representations of external (Se) and internal (Si) source and sink terms. Each rectangular box in Fig. 6.1 represents a component being simulated by the model. The arrows between components represent the biochemical transformation of one substance to the other. An arrow with one end unattached to a component (rectangular box) represents an internal source (or sink) due to the biochemical reaction or an external source (or sink) (Khodadadi Darban 2010).

Fig. 6.1 Schematic diagram of interacting water quality state variable (Khodadadi Darban 2010)

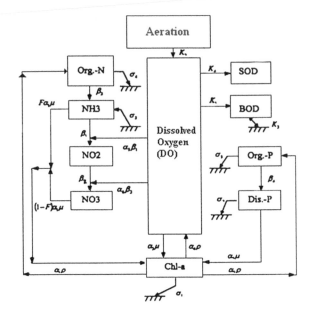

6.3.1 Boundary Conditions

Boundary conditions need to be specified in the water quality model at four boundaries: the free surface, bottom, upstream and downstream boundaries. At the free surface, the wind-induced DO aeration is incorporated into the model using the equation developed by Banks and Herrera (1977).

There is no other mass flux through the free surface. The mass fluxes at the bottom are specified specifically for each state variable by settling and benthic fluxes. The contributions of non-point source loadings from the upstream drainage area are specified as mass fluxes at the upstream boundary for each of the state variables (Khodadadi Darban 2010).

The surface elevation is specified as a function of time either with harmonic functions or with time-series data measured at this boundary. In calculating velocities at the open boundary, the horizontal velocities are linearly extrapolated to a fictitious model transecting outside the estuarine mouth, and the advective and diffusive terms are calculated over this fictitious model segment.

6.4 Study Area

6.4.1 The First Case Study

Karkheh dam is located 40 km far from West of Andimeshk city and 160 km far from Ahwaz city. Figure 6.2 shows the location of the Dam. The main objective of

6.4 Study Area

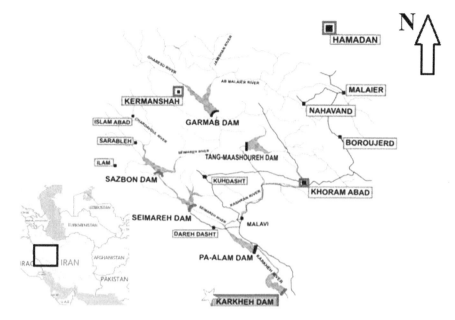

Fig. 6.2 Situation of Karkheh Dam

construction of the dam was to supply water for irrigation of Abbas, Hmidiyeh, Ghods and Azadegan plains, for generating electricity and to control the seasonal flooding.

In this case study, we monitored TDS and Temperature parameters in different water depth in the axes of the dam for 8 months, and then applied the two dimensional Ce-Qual-W2 model to the data and with secondary data from different Iranian organizations, (Meteorological Organization, and Power Ministry) used to develop a suitable model with which it is possible predict the variation of TDS in the future.

Data was collected at three stations (Jologeir, inside the reservoir and Payapol stations) which are shown in Fig. 6.2.

Considering all conditions and constraints governed in this study, it was decided for experimental work to be carried out during stratification of the reservoir. All experimental work was carried out in 2003. For water quality simulation in the Karkheh reservoir, it was necessary to have some the following information (Barderghasemi 2002):

1. Reservoir Geometry
2. Inlet and outlet discharge
3. Reservoir water level
4. Meteorology data (air temperature, speed and direction of wind, dew point, cloud cover)
5. Water quality parameters (water temperature, TDS)

Item 1–4 was gathered from Power Ministry and Meteorological Organization data and item 5 were assessed monthly from May 2003 to December 2003 by Badrghasemi (2002) for two reasons which are as follows:

Collection of water quality data which is necessary for Ce-Qual-W2 model calibration and confirmation including water temperature and TDS parameters.

Confirmation of two dimensional assumption of the reservoir.

Data of Jologeir and Payapol stations was applied as boundary conditions and data from inside the reservoir was used for calibration and confirmation of the model (temperature profile and TDS of May 2003 were used as initial conditions of the model).

Considering the shape of the reservoir which is 65 km long and 5 km widths and variation of TDS, temperature parameter at the surface of the water was about 3–5 %. It was assumed that the variation of water quality across the width of the reservoir was negligible. Therefore, for simulation of TDS and Temperature a two dimensional model (Ce-Qual-w2), was considered suitable for the evaluation of the reservoir.

Another reason for selecting the two dimensional model was that the distance between two outlets of the reservoir which was relatively high (Dashte Abbas tunnel in 40 km and outlet of the reservoir in 65 km) and variations in length of the reservoir are visible.

Ce-Qual-W2 is a two-dimensional model which was applied to simulation of hydrodynamic and water quality of the reservoir in length and height, it was used for all mean parameters in the width of a reservoir.

The selected model in this case study was carried out for 688 days of simulation, and the data of TDS and temperature of the first 200 days of simulation were used for confirmation of the model and the rest of the data used for prediction of variation of water quality in the reservoir. The first day of experimental work was Mays 5th 2003; this was the day also selected as the first day of simulation.

The length of the reservoir was divided into 66 equal parts, each of them equal to 1,000 m and the height of the reservoir was divided into 62 layers, the height of each layer was between 1.5 and 4 m. Figure 6.3 indicates a simulation of the geometry of the reservoir.

First, the reservoir geometry defined for the model, second, meteorological data file, initial conditions, boundary condition and wind confinement correction defined.

The first step of constructing a model is model calibration and also Ce-Qual-W2 model needs to evaluate the temperature and the amounts of TDS of each layer for analyzing densities of different layers. Hence, before the simulation of TDS, it is necessary to construct temperature stratification models and then it is necessary to calibrate temperature and TDS parameters. Whenever the results of simulation are confident, then the prediction of TDS parameter will be carried out for future.

The result for the first temperature and TDS simulation models were not matched to the experimental study. Therefore, the calibration of the model begun. First, the geometry and second the water level of the reservoir was calibrated.

6.4 Study Area

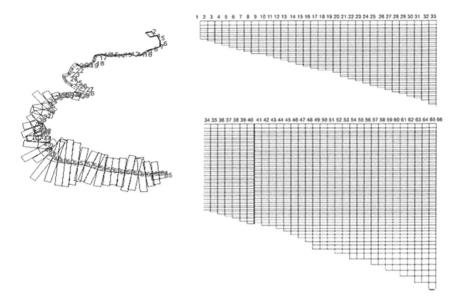

Fig. 6.3 Reservoir geometry simulation (Badrghasemi 2002)

The maximum percentage of the difference between actual and model evaluated volumes of the reservoir was about 0.5 % and the maximum percentage of the difference between actual water levels and the model evaluation in the reservoir was about 0.61 m. Figure 6.4 indicates a comparison between actual and model of water level from May to December 2003.

After calibration of hydrodynamic characteristics of the reservoir, the water quality of the reservoir was calibrated. Results of model sensitivity analysis are as follows: Vertical distribution coefficient AZ (man) had affected the model result when analysis was carried out in a vertical direction. Increasing the mentioned coefficient in the model, lead to breaking thermo cline in depth and also surface absorption radiation coefficient in depth affected the construction of the model and the use of a bigger coefficient led to a simulation result closer to reality.

Cloud cover coefficient also had a significant effect in the reduction of solar radiation and correction of it caused an improvement in the result. The wind velocity correction was an important parameter for model calibration.

Figures 6.5, 6.6, 6.7 and 6.8 describe the comparison between results of simulation of temperature and TDS profiles with experimental work which was carried out from May to December 2003.

As presented in Fig. 6.5 in late June and early July 2003, water temperature and TDS on the top of the water level were about 32 °C and 370 mg/l respectively; and in the bottom of the reservoir temperature and TDS was 14 °C and 680 TDS mg/l respectively. According to the results, there was stratification, also the inlet water to the reservoir had a water temperature about 20 °C and TDS 690 mg/l.

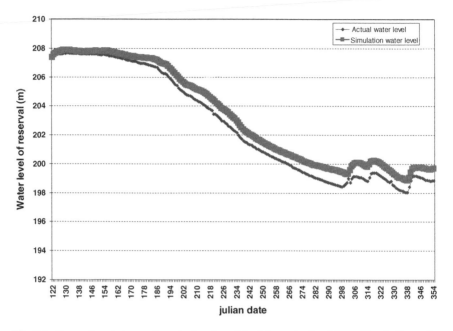

Fig. 6.4 Comparison between the actual and model of the water level from May to December 2003 (Badrghasemi 2002)

In late June and early July 2003, weather conditions relating to May and early June were hot.

The water temperature at the inlet increased up to 28.5 °C from July to August. The increase of TDS in the inlet to 1,033 mg/l in August caused an increase in the TDS gradient in Termocline in the same month. At the end of September, the depth of Epilimnion gradually increased because the weather was cold. Thus, the stratification of the reservoir was changed to be homogeneous in it's depth. In addition, because of the decrease of the water temperature and consequently the increase of inlet water density, the newly inlet water flow will penetrate the lower layers of water relating to last month and the temperature increased to 21° in November. In addition, the TDS in inlet water decreased to 770 mg/l. At this time the surface water temperature will be 26 °C and the amount of TDS was 500 mg/l. In December, because of the decrease in air temperature, the inlet water density was increased and thus, the water tends to move to the deepest layers of the reservoir and the Hipolimnion layer direction. During this period, the water temperature was 14 °C.

In January a further reduction in air temperature and the increased sun shine leads to an increase of the Epilimnion layer. At the end of January, the thickness of upper and bottom layers increases and the Termocline layer is broken. In February, the amount of TDS in the surface layers is a little bit more than in the bottom layers. On this month, the changes in TDS density in the lower layers of the reservoir, which the input water tends to move, depend on the TDS of the input water.

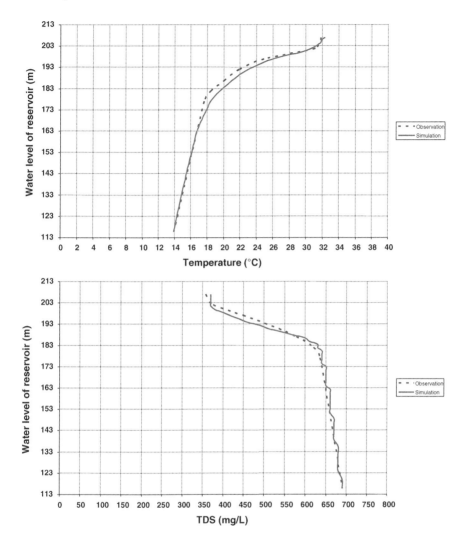

Fig. 6.5 Comparison between actual and model of TDS and temperature—July 2003 (Badrghasemi 2002)

In March, the air temperature gradually increases and the density of entering water decreases minimally, so that the input water tends to move in the upper layers of the reservoir. In addition, because of an increase in the surface temperature of the reservoir, the situation to further layer is provided. During this month, the TDS of input water decreases and the TDS levels in the upper layers of the reservoir are attenuated, but in the lower layers the TDS density remains unchanged. This layering cycle is repeated each year. In the event of a flood in the first months of a year, the cycle will be interrupted for a short period before returning to the normal.

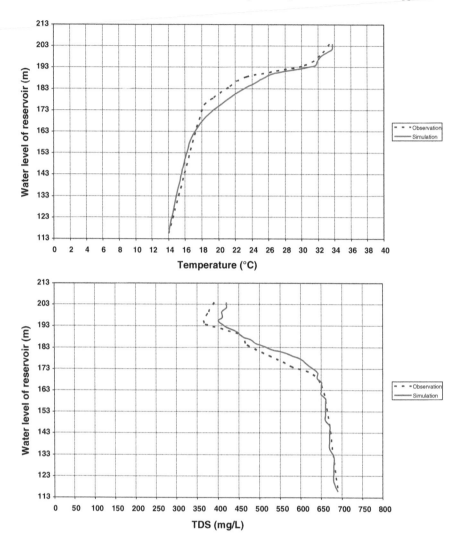

Fig. 6.6 Comparison between actual and model of TDS and Temperature—August 2003 (Badrghasemi 2002)

Figures 6.9, 6.10 and 6.11 indicates prediction results from the Simulation Models of Temperature and TDS in January, February and March 2003. However, real data for confirmation was not available.

Also Fig. 6.12 presented the comparison between the amount of the TDS in the inlet and outlet of the Karkheh reservoir from 2003 to 2004.

6.4 Study Area

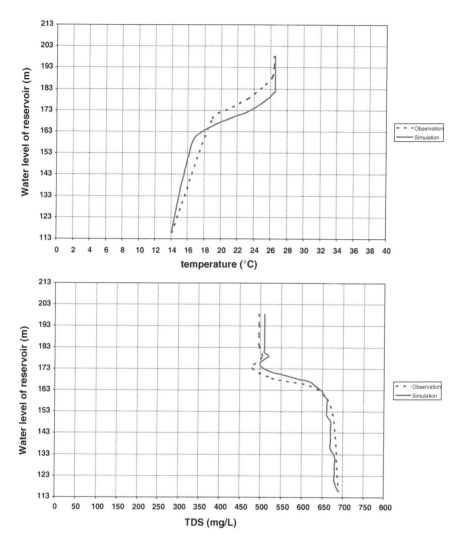

Fig. 6.7 Comparison between actual and model of TDS and Temperature—November 2003 (Badrghasemi 2002)

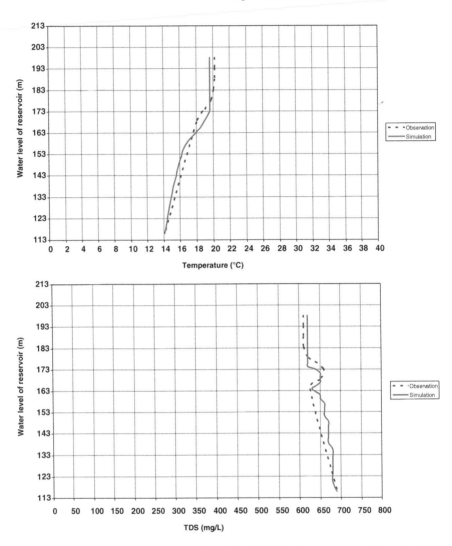

Fig. 6.8 Comparison between actual and model of TDS and temperature—December 2003 (Badrghasemi 2002)

6.4 Study Area

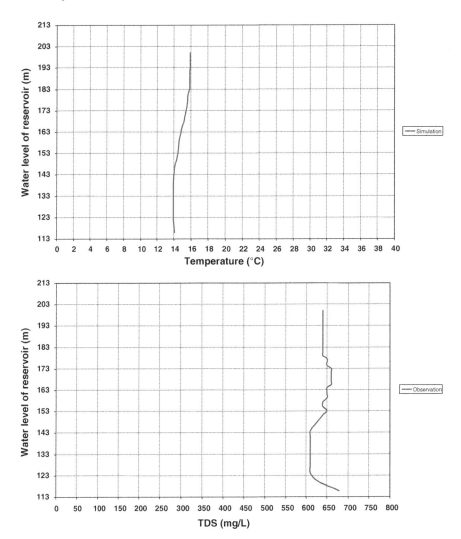

Fig. 6.9 Simulation Models of Temperature and TDS—January 2003 (Badrghasemi 2002)

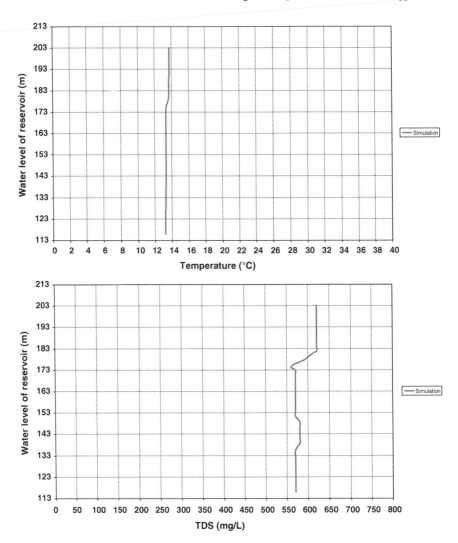

Fig. 6.10 Simulation models of temperature and TDS—February 2003 (Badrghasemi 2002)

6.4 Study Area

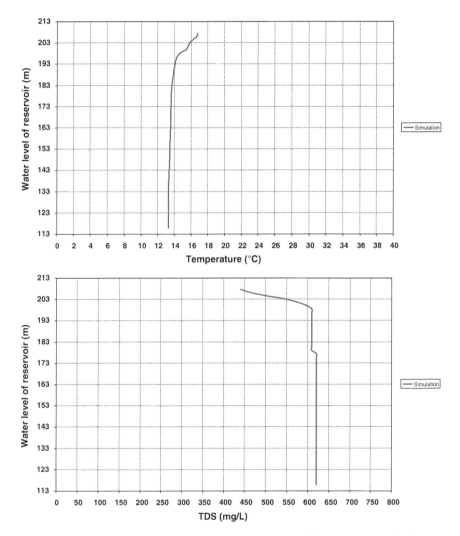

Fig. 6.11 Simulation models of temperature and TDS—March 2003 (Badrghasemi 2002)

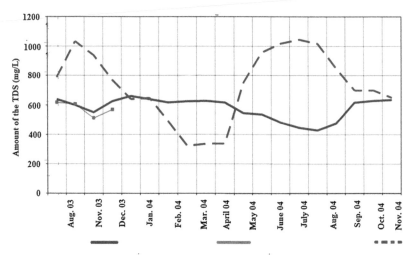

Fig. 6.12 The comparison between the amount of the TDS in the inlet and outlet of the Karkheh reservoir from 2003 to 2004 (Badrghasemi 2002)

6.4.1.1 Comment

As a whole, CE-QUAL-W2 model is a suitable tool for water quality prediction on in reservoir. The results of the simulation in this case study showed that the existing water has a good quality inside the whole of the reservoir and in 2003; there was no accumulation of TDS at any points in the Dam. It is necessary to propose different theories to estimate the quality of possible brininess of Karkhe reservoir.

In the simulation, it was assumed that with the increase of the discharge of 2004 relating 2003, the inlet water to the reservoir increased, and the amount of TDS in 2004 was equal to TDS in 2003. With this assumption, the simulation results indicated that TDS in the reservoir in March 2004 was 17 mg/l more than in March 2003. Considering these results if the circumstances do not change during the next few years, the TDS in Karkhe reservoir may increase by 17 mg/l each year. Taking the projected increase of TDS into account it is therefore possible to identify the timepoint at which TDS pollution of the reservoir water will render it unsuitable for irrigation.

Considering the EPA standard, and also above preconception, it will take 24 years for the amount of TDS in the Karkhe reservoir to reach 1,000 mg/l from a starting point of 600 mg/l in 2003. In other words, the Karkheh may not be suitable for agricultural use in 24 years.

The mentioned theory depends on gaining the accurate information about changing of TDS in Karkhe reservoir. If, as we project, the amounts of TDS increase each year, the assumption will be reasonable.

As presented in Fig. 6.12 when inlet water has a high amount of TDS, the outlet water has TDS. Hence, the water in the reservoir will gradually become salty.

6.4 Study Area

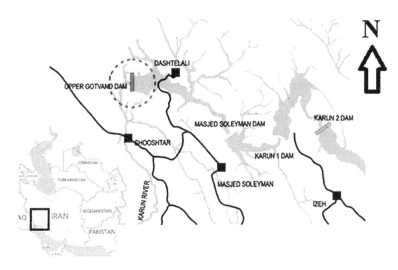

Fig. 6.13 Map of study area in Kārūn River

This graph shows inequality between the inlet and output TDS. The inlet TDS is usually more than the outlet TDS (Badrghasemi 2002).

This study indicates that a TDS buildup is to be expected in the Kharkheh reservoir in the coming years and that this will limit the use of the water for irrigation purposes. Thus, we concluded that the implementation of TDS accumulation prevention strategies is necessary to ensure long-term usage of the reservoir.

6.4.2 Second Case Study

The Kārūn River is the most important river in the south of Iran. However, it receives various wastewaters from a wide range of sources, including domestic, agricultural and industrial. Thus, analyzing the effects of pollutants on the water quality of the Kārūn River is inevitable. Studies have been conducted in other parts of this river, or similar rivers in Iran but a lack of analysis on this lake persists.

The study area is the region upstream of the Gotvand Dam on the Kārūn River which is 32,425 km^2 to well beyond Gotvand Dam and two main cities following it. The Gotvand Dam is located in the south west part of Iran in the Khuzestan province between 48°49′ to 48°57′E and 32°12′ to 32°17′N.

In this case study the QUAL2K was applied, a numerical hydrodynamic and water quality model provided by the U.S. EPA, to analyze the effects of pollutants on the water quality of the Kārūn River in the Gotvand-Shooshtar region. The Kārūn River is located in South West of Iran and many dams are constructed and are being constructed on this river for the generation of electricity and water consumption. At the same time the Kārūn River extends through a large part of the south–west thus receiving lots of wastewater from different sources. The water quality model is used to simulate the water quality condition in the Kārūn River, estimating the minimum in stream flow for ecosystem survival especially fish survival and propose management strategies to improve the water quality (Khodadadi Darban 2010).

In this work we focused on the region upstream of the Gotvand dam, to the Dam and two cities which are located downstream of the dam. The reason mainly being that the Gotvand dam is the last major dam in the Kārūn River, so all the untreated waste from upstream and the many pollutants in this region make this segment of the river critical for analysis, and the reduction in flow after the Gotvand dam makes analyzing its effects on the water quality downstream of the dam (Fig. 6.13).

During low-flow periods the in stream flow becomes minimal and the Dissolved Oxygen (DO) declines. In this event, the biological habitat of the river is endangered, fish deaths happen in low DO concentrations, and thus studying the DO is of importance for this research (Jöhnk and Umlauf 2001). The Gotvand Dam reservoir is also used for municipal and agricultural water supplies, making water quality studies in this part inevitable (Lindenschmidt et al. 2004).

6.4.2.1 Model Calibration and Verification

Reliable simulation can be undertaken for management planning by calibrating and verifying the model. A water quality model needs to be calibrated and verified with respect to the prototype conditions of the water body to which it is applied. The model was calibrated by hydraulic constants and coefficients provided by Dez Ab for 1999 and coefficient values from calibration conducted by the Water Resources Management Company of Iran in 2001. The model was then recalibrated and verified with respect to the field data taken in 2004–2005 for the Kārūn River (Khodadadi Daran 2010).

The recalibration is far more difficult for the water quality model than for the hydrodynamic model, due to the large number of water quality state variables and biochemical reaction coefficients involved. Since the model predictions will change depending upon the selection of the values of biochemical coefficients, consistent coefficient values should be used for different simulation runs. That is, the coefficient values should be transferable for the model predictions to compare with independent sets of field observations (Snowling and Kramer 2001, Wells 2001).

The model was run for the year 2005. The model results for daily average concentrations at the surface and bottom layers was compared with the observed values at the corresponding stations. The model results and field measurements are provided in the respective figures, they indicate that the model results and field measurements are generally in good agreement, with errors of less than 6.5 %.

The temporal variability of the model results is generally different from those of the field data, because the model results represent the daily average values of the lateral average concentrations while the field data were point measurements, and also because of the random variability inherent to a natural system, thus errors to this extent were allowable.

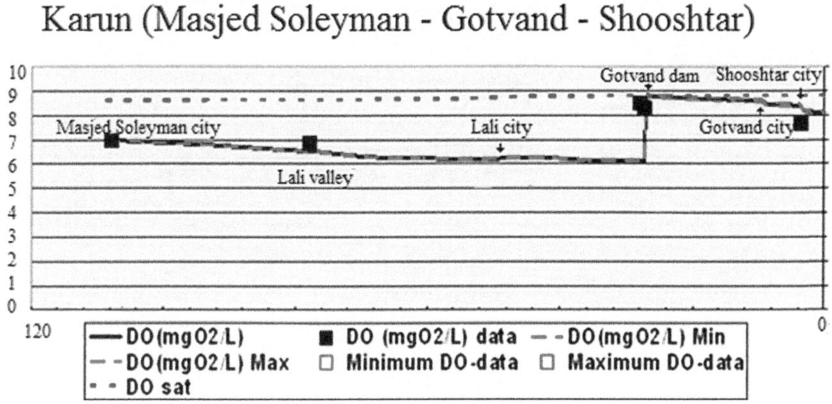

Karun (Masjed Soleyman - Gotvand - Shooshtar)

Fig. 6.14 Dissolved oxygen (DO) in the study area during normal conditions (mg/l)

The errors averaged over all stations were 0.5 and −0.3 mg/l, respectively, for CBOD, and DO. The negative values indicate that the measured data are higher than the model results.

Analyzing the values of Dissolved Oxygen in normal conditions in the river (Fig. 6.13) indicated a general decrease, as a result of agricultural, industrial and municipal polluting sources, and then a sudden peak resulting from the Gotvand dam. A decrease was again witnessed after the dam from agricultural sources and two municipal sources. The lowest the DO reaches was 6 mg/l which is acceptable, but could endanger aquatic life in the early stages of life based the Dissolved Oxygen Levels for Aquatic Life in Table 6.1. Fish deaths have occurred in the Kārūn River in recent years, causing a stench and raising concerns. The main streams that fish kills happen are in cities that high volumes of wastewater enter the river, especially in the region of Ahvaz City. The fish in these regions if not dead are is not edible due to the pollution in the river.

Investigation of fish deaths by researchers has shown that low Dissolved Oxygen (DO) was the main reason for this problem to occur (Meyer and Barclay 1990).

Oxygen is essential for fish to survive. Fish deaths will occur if oxygen levels drop below a certain critical concentration that depends on the fish species. For example Gheze Ala fish can live when the DO concentration in the river is maintained above 5 mg/L (Fig. 6.14).

Figure 6.15 presents the changes in BOD during normal conditions in the Kārūn River. The main locations affecting the quality are labeled in the figure. The BOD changes in the stream are affected as a result of the receiving pollutants; however, due to the relatively high water flow the changes are not considerable and the rivers self-restoration is affecting the quality more and is creating an acceptable water quality with a BOD of less than 3 mg/l, which is acceptable by Iran's standards based on Table 6.2 and most International Water Quality standards.

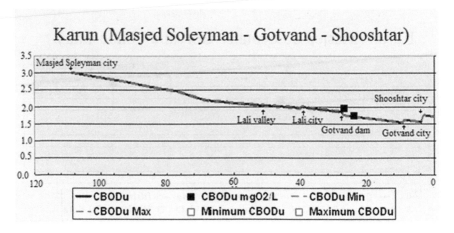

Fig. 6.15 BOD in the study area during normal conditions (mg/l)

Table 6.2 Dissolved oxygen level (mg/l) for aquatic life (mg/l); Quality Criteria for Water (1986) (EPA 440/5-86-001)

	Warm water conditions		Cold water conditions	
	Other stages of life	Early stages of life	Other stages of life	Early stages of life
30 day average	5.5	–	6.5	–
7 day average	–	6	–	(6.5) 9.5
Minimum 7 day average	4.0	–	5.0	–
1 day minimum	3.0	5.0	4.0	(5.5) 8.5

Studying the water quality of the river under critical conditions is then considered as the quality is not affected much during normal conditions. Figure 6.15 presents the changes in DO in low flow periods. The reduction in flow greatly affects the Dissolved Oxygen variations due to incoming pollutants, and a lower DO with greater variations is noticeable. Based on Table 6.2 aquatic life at early and other stages of life is endangered in the cold season as DO falls to about 6 mg/l. However, it passes Iran's criteria for water quality compared to Table 6.3.

Figure 6.16 illustrates BOD variations in the Kārūn River during low flow periods. BOD variations are also more considerable in low flow periods and the changes due to receiving pollutants are much greater than in the normal flow as shown in Fig. 6.16. Comparing the values to General Criteria for Water Quality in Iran (Table 6.3) and EPA Dissolved Oxygen Levels for Aquatic Life (Table 6.2) again proves the BOD is in agreement with the current standards and is not considered critical.

The daily freshwater discharge in the Kārūn River during 2005 was higher than in many other years and may not be representative of the danger to fish survival during

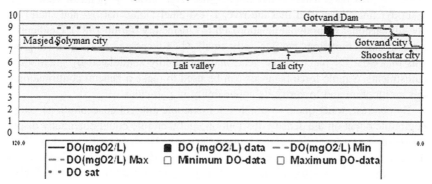

Fig. 6.16 DO variations in Kārūn River during low flow periods (mg/l)

Table 6.3 General criteria for water quality in Iran; Iran's Department of Environment (DOE) (2003)

DO (mg/l)	BOD_5 (mg/l)
5	5

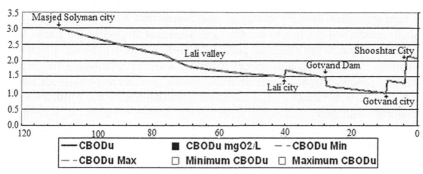

Fig. 6.17 BOD variations in Kārūn River during low flow periods (mg/l)

drought periods. Therefore, a designed flow must be maintained during the construction of the Gotvand Dam and after Operation for the keeping the eco-system safe for aquatic lives. Based on the results of this study as indicated in the Figs. 6.16 and 6.17, a flow of 90 m^3/s must be maintained in the river. This design flow may be predicted higher if the effects of pollutants sources in the entire Kārūn region in the downstream of the Gotvand Dam are considered (Khodadadi Darban 2010).

However, it may be questioned to what extent a one-dimensional model adequately represents a natural system such as a river. From the point of view of the hydrodynamics, provided that the river is extremely long, narrow and shallow, a one-dimensional hydrodynamic approach is satisfactory (Orlob 1983). Gradients in the horizontal directions are generally small when compared to the vertical gradients that exist for much of the year, and are rapidly annihilated by gravitational adjustments. Simple force balances can be used to make a general confirmation of this assumption and to verify its applicability in particular.

By contrast, water quality variables, for example nutrient concentrations, exert a negligible effect on the density distribution, and therefore could potentially display a two- or three-dimensional distribution despite a 1D density distribution. While this is recognized as a shortcoming of a 1D model, it does not necessarily imply that a multi-dimensional approach would produce a more correct picture. Indeed, given the difficulty of setting realistic initial conditions for all water quality variables in a multi-dimensional model and the difficulty of knowing all of the input fluxes at the spatial scale of the model, a multi-dimensional predictive capacity (as opposed to a multi-dimensional verification capacity) is a highly uncertain outcome in many natural systems. By explicitly recognizing that the output from a 1D model is a horizontally averaged result, the 1D assumption provides a base level prediction which can be achieved with a greater degree of certainty and from which inferences about possible horizontal distributions can be drawn (Khodadadi Darban 2010).

6.4.2.2 Comment

A time-dependent, laterally averaged, one-dimensional hydrodynamic and water quality model was applied to the Gotvand-Shooshtar Region of the Kārūn River system to simulate the effects of pollutants on dissolved oxygen and CBOD distributions. The water quality model, supplied with the information for physical transport processes from the hydrodynamic model, provides real-time predictions of water quality state variables. Hydrodynamic model calibration and verification were conducted with mean data range, time varying surface elevation and longitudinal velocity, turbulent mixing, and salinity distribution in the Kārūn River system. The model was updated with the geometric data for 2005 and recalibrated with field data for water surface elevation and velocity measured in the same year. The overall performance of the model was in qualitative agreement with field data. The water quality model was also recalibrated using field data collected in 2005. Considering the random variability inherent to natural systems and the goal of consistency in calibrated coefficients, the agreement between the model predictions and field observations is more than satisfactory.

The recalibrated model was used to perform sensitivity analyses. It is demonstrated that the DO concentration in the river is very sensitive to the magnitude of river flow, particularly during the low-flow period. Increasing river flow significantly raises the DO level. The model is then used to simulate various water quality management strategies including river flow management and wastewater loading reduction. The Gotvand Dam may impound water during the high flow periods and release freshwater to maintain the required instream flow during the drought periods (Khodadadi Darban 2010).

References

Badrghasemi B (2002) TDS study for Karkheh Dam using Ce-Qual-W2, M.Sc. thesis, Science and Technology University, Tehran, Iran (In Persian)
Banks RB, Herrera FF (1977) Effect of wind and rain on surface reaeration. J Environ Eng Div 103 (3):489–504
Beecher HA (1990) Standards for instream flows. Rivers 1(2):97–109
Berger C, Wells S (2001) The spoken river-Long lake system model, Department of Civil and Environmental engineering Portland State University and ACOE WES
Boorman DB (2003) Climate, hydrochemistry and economics of surface-water systems (CHESS): adding a European dimension to the catchment modelling experience developed under LOIS. Sci Total Environ 314:411–437
Brown LC, Barnwell TO (1987) The enhanced stream water quality models QUAL2E and QUAL2E-UNCAS: document and user manual. US EPA Office of Research and Development
Canfield DE Jr, Bachmann RW (1981) Prediction of total phosphorus concentrations, chlorophyll a, and Secchi depths in natural and artificial lakes. Can J Fish Aquat Sci 38(4):414–423
Chapra S, Pelletier G (2003) QUAL2K: a modeling framework for simulating river and stream water quality: documentation and user's manual. Civil and Environmental Engineering Department, Tufts University, Medford
Cox B (2003) A review of currently available in-stream water-quality models and their applicability for simulating dissolved oxygen in lowland rivers. Sci Total Environ 314:335–377
Deksissa T (2004) Dynamic integrated modelling of basic water quality and fate and effect of organic contaminants in rivers. Ph.D thesis, Ghent University, Belgium
Department of Environment (The DOE) (2003) Regulation and Standard of Environment (IRAN). Dayeh Sabz, Iran (In Persian)
EPA (1986) Quality criteria for water. EPA 440/5-86-001. US Environmental Protection Agency, Office of Water, Regulations and Standards, Washington, DC
Institute of Standards and Industrial Research of Iran (ISIRI) (2007) Standard for drinking water, No. 1053, 4th edn. ISIRI, Iran (In Persian)
Joehnk K, Umlauf L (2001) Modelling the metalimnetic oxygen minimum in a medium sized alpine lake. Ecol Model 136(1):67–80
Khodadadi Darban A (2010) Application of CE-QUAL-K to define water quality of Ghotvand Dam in Kaurn River. MSc thesis, University of Tehran, Iran
Lindenschmidt K-E, Eckhardt S, Wodrich R, Eckert U, Baborowski M, Guhr H (2004) Water quality modeling of a lock-and-weir system on the lower Saale River. Gas Wasserfach Wasser Abwasser 145(9):612–621
Meyer FP, Barclay LA (1990) Field manual for the investigation of fish kills
Orlob GT (1983) Mathematical modeling of water quality: streams, lakes, and reservoirs, vol 12. Wiley, Chichester
Razdar B, Mohamadi K, Samani J, Pirooz B (2011) Determining the best water quality for the rivers in the north of Iran (case study: Paksikhan River). Comput Methods Civil Eng 2:109–121
Rounds S (2001) Modeling water quality in the Tualatin River: achievements and limitations. In: Warwick JJ (ed) AWRA annual spring specialty conference proceedings, "Water Quality

Monitoring and Modeling". American Water Resources Association, Middle-burg, Virginia, TPS-01-1:115-120

Sarang A, Tajreshi M (2001) Simulation of Bookan Dam. Office of Water and Environment in Civil Engineering Faculty, Sharif University of Technology, Iran (In Persian)

Shiati K (1996) Behavior of salinity in Dams in Iran. Applied Research Planning Ministry of Power, Iran (In Persian)

Snowling S, Kramer J (2001) Evaluating modelling uncertainty for model selection. Ecol Model 138(1):17–30

Tebbutt THY (1997) Principles of water quality control. Butterworth-Heinemann, Oxford

Wells S (2001) CE-QUAL-W2 version 3 user manual appendix A: hydrodynamics and transport, May 26, 2001. http://www.ce.pdx.edu/~scott/w2. Accessed 25 Feb 2005

CPSIA information can be obtained at www.ICGtesting.com
Printed in the USA
LVOW01s1917031014

407172LV00002B/16/P

9 783662 447240